Stefan Pölz

Personalised body counter calibration using anthropometric parameters

Personalised body counter calibration using anthropometric parameters

by
Stefan Pölz

Dissertation, Karlsruher Institut für Technologie (KIT)
Fakultät für Elektrotechnik und Informationstechnik
Tag der mündlichen Prüfung: 04. Februar 2014
Referent: Prof. Dr.-Ing. habil. Manfred Urban
Korreferenten: Prof. Dr. rer.nat. Dr. h.c. Manfred Thumm
Priv.-Doz. Dr. rer.nat. Bastian Breustedt

Impressum

Scientific
Publishing

Karlsruher Institut für Technologie (KIT)
KIT Scientific Publishing
Straße am Forum 2
D-76131 Karlsruhe

KIT Scientific Publishing is a registered trademark of Karlsruhe
Institute of Technology. Reprint using the book cover is not allowed.

www.ksp.kit.edu

Print on Demand 2014

ISBN 978-3-7315-0174-9
DOI: 10.5445/KSP/1000038415

Personalised body counter calibration using anthropometric parameters

Zur Erlangung des akademischen Grades eines

DOKTOR-INGENIEURS

von der Fakultät für
Elektrotechnik und Informationstechnik
des Karlsruher Instituts für Technologie (KIT)
genehmigte

DISSERTATION

von

Dipl.-Inform. Stefan Pölz
geb. in Erfurt

Tag der mündlichen
Prüfung: 04. Februar 2014
Hauptreferent: Prof. Dr.-Ing. habil. Manfred Urban
Korreferenten: Prof. Dr. rer.nat. Dr. h.c. Manfred Thumm
Priv.-Doz. Dr. rer.nat. Bastian Breustedt

Zusammenfassung

Die Aktivitätsbestimmung mit Teilkörperzählern ist ein *In-vivo*-Messverfahren zur Überwachung von Personen mit erhöhtem Risiko einer Radionuklidinkorporation. Dabei werden Strahlendetektoren relativ zum Körper angeordnet, um Depositionen von Radionukliden in anatomischen Strukturen, wie Lungen, Leber oder Knochen, zu quantifizieren. Dieses Verfahren hängt vom spezifischen Messsystem ab und ist sensitiv bezüglich der individuellen Anatomie der zu messenden Person. Die gemessenen Aktivitäten sind die Basis für eine anschließende Dosisabschätzung und setzen eine aufwändige Kalibrierung des Messsystems voraus. Die Kalibrierung involviert typischerweise experimentelle Messungen an anthropomorphen Phantomen in Standardmessanordnungen. Aktuell eingesetzte Kalibrierverfahren erlauben zusätzlich eine Personalisierung speziell für Lungen- und Lebermessungen abhängig von Körpergewicht und -größe in Bezug auf ein konfigurierbares Referenzphantom.

In dieser Arbeit werden die aktuell eingesetzten Personalisierungsmethoden mit Hilfe von Strahlentransportsimulationen und Computerphantomen aus medizinischen Bilddaten revidiert und erweitert. Das entwickelte Verfahren erlaubt die Berechnung von Kalibrierfaktoren in beliebigen Messanordnungen und von anthropometrischen Parametern zur Quantifizierung der individuellen anatomischen Eigenschaften. Diese Werte dienen der statistischen Analyse und Erstellung von Schätzern, die personalisierte Kalibrierfaktoren aus personenspezifischen Werten von anthropometrischen Parametern ableiten. Dieser systematische Ansatz liefert bessere Schätzwerte für die Detektorkalibrierung, die sich unmittelbar auf die Aktivitätsbestimmung in den betroffenen Strukturen und die Dosisabschätzung für das Individuum auswirken.

Das Verfahren wurde in Form eines abstrakten, modularen Datenmodells und eines Softwarewerkzeugs zur Modellierung, Simulation, und Auswertung von allgemeinen Messszenarien, und einer Methode zur statistischen Analyse des Zusammenhangs von anthropometrischen Parametern und Kalibrierfaktoren implementiert. Das erlaubt eine effiziente und reproduzierbare Modellierung zur virtuellen Rekonstruktion von Messungen und zur Durchführung von Sensitivitätsanalysen. Das Verfahren wurde zur Kalibrierung des In-Vivo-Messlabors (IVM) des Karlsruher Instituts für Technologie (KIT) angewendet, das aus vier frei positionierbaren Reinstgermaniumdetektoren mit Standardmessanordnungen für Lungen, Leber, Knie, und Kopf besteht. Aufgrund der interindividuellen anatomischen Variationen in den verwendeten Phantomen und zusätzlicher Sensitivitätsanalysen war es möglich Abschätzungen zu den erwarteten Unsicherheiten anzugeben und durch ein algorithmisch reproduzierbares Vorgehen bei der Kalibrierung zu reduzieren.

Abstract

Body counting is a method for *in vivo* activity assessment applied to the monitoring of people with high risk of radionuclide incorporation. Energy-sensitive radiation detectors are arranged relative to the body to quantify radionuclide deposits in anatomical structures, such as lungs, liver and skeleton. This method depends on the specific detection system and is sensitive to the individual anatomy of the person. Accurate activity estimates, which are the basis for dose calculation, require extensive calibration procedures typically involving experimental measurements of anthropomorphic phantoms conforming to a reference person. Current calibration methods offer personalisation for lung and liver counting only with respect to body mass and height and do not specify uncertainties.

This work revises and extends the currently applied personalisation methods using radiation transport simulation in combination with computational phantoms derived from medical imaging data. A framework was developed that allows computation of samples of calibration factors for various anatomies in standard measurement setups and anthropometric parameters quantifying anatomic properties. Those samples are applied to create statistical models to derive personalised calibration factors given specific values of anthropometric parameters measured on the person. This gives better estimates in activity assessment and, thereby, dose calculation while quantifying and reducing uncertainties.

The framework was implemented in form of an abstract, modular data model, a software tool for modelling, simulation and evaluation of general body counting scenarios, and a statistical analysis method for correlating anthropometric parameters and calibration factors. This allows efficient and reproducible modelling for virtual measurement reconstruction as well as sensitivity analyses. The framework was applied to the calibration of the In Vivo Measurement Laboratory (IVM) at Karlsruhe Institute of Technology (KIT) comprising four freely arrangeable high-purity germanium detectors in lung, liver, knee and head measurement setups. Because of the interindividual anatomical variations in the applied phantoms and additional sensitivity analyses, it was possible to give estimates of the expected uncertainties and to reduce them through an algorithmically reproducible approach on calibration.

Acknowledgements

First of all, I would like to thank my supervisors Prof. Manfred Urban and PD Dr. Bastian Breustedt for guiding me in the past three years and providing valuable feedback for this work.

My special thanks to Dr. Lars Hegenbart for introducing me to the topic of radiation protection research, creating the idea and initial version of VOXEL2MCNP and providing ideas for its further development. I am grateful to Marco A. Harrendorf, Sven Laubersheimer, Thomas Keck, Andreas Benzler and, especially, Jakob S. Eberhardt, who developed and tested parts of the software. I also appreciate the contributions of Thomas Schneider and Sven Laubersheimer as part of their diploma and bachelor theses.

I would like to thank Dr. W. Paul Segars of Duke University, Prof. George X. Xu of Rensselaer Polytechnic Institute, Prof. Richard Kramer of University of Pernambuco, Prof. Wesley E. Bolch of University of Florida and Dr. Choonsik Lee of National Cancer Institute for providing computational phantoms that have been applied throughout this work.

I also would like to thank Rodolfo Cruz Suaréz and Josef Heiss of International Atomic Energy Agency for providing the JAERI phantom for measurements, Dr. Schober and his associates of the German Cancer Research Center for providing the LLNL phantom lung set and for handling the radioactive materials, and Dr. Jürgen Wilbert of the University Hospital Würzburg for providing respiratory-correlated computed tomography data sets.

I thank all my colleagues at the Institute for Nuclear Waste Disposal and the former Institute for Radiation Research for welcoming me and helping me with unfamiliar topics.

Finally, I would like to thank the Federal Ministry of Education and Research who made this work possible with their funding of the project "Strahlung und Umwelt 2" (02NUK015A) and all of the collaboration partners involved in the project, and, in particular, Dr. Angelika Bohnstedt.

Contents

Figures

Tables

Listings

Abbreviations

AF	Adult female
AM	Adult male
BMI	Body mass index
CWT	Chest wall thickness
ENSDF	Evaluated Nuclear Structure Data File
FASH	Female adult mesh
FWHM	Full-width-at-half-maximum
HPGe	High-purity germanium
IAEA	International Atomic Energy Agency
ICRP	International Commission on Radiological Protection
ICRU	International Commission on Radiation Units & Measurements
IVM	In Vivo Measurement Laboratory / In-vivo Messlabor
JAERI	Japan Atomic Energy Research Institute
KIT	Karlsruhe Institute of Technology
LLNL	Lawrence Livermore National Laboratories
MASH	Male adult mesh
MCNPX	Monte Carlo N-Particle Extended
NHANES	National Health and Nutrition Examination Survey
NURBS	Non-uniform rational basis spline
RPI	Rensselaer Polytechnic Institute
UFHADF	University of Florida hybrid adult female
UFHADM	University of Florida hybrid adult male
V2M	Voxel2MCNP
XCAT	Extended cardiac-torso
XML	Extensible markup language

Symbols

Symbol	Name	SI unit
α	Alpha particle	
β^-	Beta-minus particle (electron)	
β^+	Beta-plus particle (positron)	
γ	Gamma quantum (photon)	
η	Counting efficiency	
ϑ	Scattering angle	rad
λ	Decay rate	Bq
λ	Kernel bandwidth	
μ	Linear attenuation coefficient	$\mathrm{m^{-1}}$
μ	Mean	
μ/ρ	Mass attenuation coefficient	$\mathrm{m^2\,kg^{-1}}$
ν_e	Electron neutrino	
$\overline{\nu}_\mathrm{e}$	Electron anti-neutrino	
ρ	Density	$\mathrm{kg\,m^{-3}}$
σ	Cross section	$\mathrm{m^2}$
σ	Standard deviation	
Φ	Fluence	$\mathrm{m^{-2}}$

Symbol	Name	SI unit
A	Activity	Bq
A	Area	m^2
c	Circumference	m
C	Number of channels	
cps	Count rate	s^{-1}
e^-	Electron	
e^+	Positron	
E	Energy	J
E_{FWHM}	Energy resolution	J
h	Body height	m
K	Kernel function	
m	Body mass	kg
N	Number (amount)	
n	Principal quantum number	
p, P	Anthropometric parameter value	
R	Risk (error)	
s, S	Source	
t	Time	s
t, T	Tally	
v	Electric potential	V
V	Volume	m^3
w	Fraction (weight or volume)	
x, X	Phantom or person	
x_{cw}	Chest wall thickness	m
y	Yield	
Z	Atomic number	

Part I

Introduction

1 Introduction

1.1 Background

While people are constantly exposed to ionizing radiation from radioactive materials present in nature, they are also at risk of exposure to man-made sources while trying to take advantage of beneficial effects of radiation — for example, as workers in agriculture, industry, medicine, nuclear fuel processing, or scientific research; as patients when receiving medical treatment or imaging in nuclear medicine, interventional radiology, or radiation therapy; or as members of the public from accidental releases of radioactive materials. The focus of this work is on radiation exposure from sources incorporated into the human body.

1.1.1 Radioactive decay

The decay of radioactive materials produces different types of radiation in a stochastic process depending on the radionuclide. The basic types of radiation are alpha, beta and gamma radiation. The expected number of decays occurring in a given quantity of medium per unit time is quantified as *activity*.

Each emission has a characteristic energy or energy spectrum. Depending on type, energy and exposure scenario, radiation can penetrate the human body and deposit energy due to its interaction with tissue, which may cause deterministic or stochastic health effects. Examples for deterministic effects are erythema, hypothyroidism, lens opacity and sterility. Stochastic effects are primarily radiation-induced cancer and hereditary effects. The amount of energy imparted in a given quantity of matter per unit mass is quantified as *absorbed dose*.

1.1.2 Incorporation pathways

Radioactive materials may be incorporated into the human body through different pathways. The body interacts with radionuclides as part of the metabolism depending on size and chemical structure of the molecule that contains the radionuclide. Airborne radionuclides can be inhaled as dust or vapour and accumulate in the respiratory system. Radionuclides in solution can be ingested and then digested by the gastrointestinal system. They may also be incorporated through direct contact with open wounds. All pathways eventually lead to the circulatory system from where the radionuclides may reach any part of the human body.

While radionuclides are transported inside the body, they deposit energy in the surrounding tissue depending on their activity and retention time in the specific structures. Tissues vary in their sensitivity with regard to radiation-induced cancer. Sensitive tissues and organs are, for example, red bone marrow, colon, lungs, stomach, breasts, and gonads (ICRP 2007).

Eventually, incorporated radionuclides may leave the body through excretion, perspiration, or exhalation. But they can have considerably large retention times, for instance, ^{210}Pb, ^{239}Pu and ^{241}Am in bone, ^{125}I and ^{131}I in thyroid, ^{137}Cs in muscle tissue, ^{239}Pu and ^{241}Am in liver, ^{235}U, ^{238}U, ^{239}Pu and ^{241}Am in lungs, or ^{239}Pu in lymph nodes (ICRP 1997).

1.1.3 Radiation protection

The goal of radiation protection is the protection of people and the environment from the harmful effects of ionizing radiation. It is of interest to the individual and the regulatory authorities to limit and monitor exposures that result from intake of radionuclides.

The International Commission on Radiological Protection recommends methodologies for exposure assessment (ICRP 2007) containing models to calculate the radiation- and tissue-dependent dose to a reference person (ICRP 2002) with a case-specific exposure. This value is used to estimate the lifetime health risk of the person. These recommendations were adopted by the International Atomic Energy Agency in their guidelines and standards for exposure assessment due to radionuclide intake (IAEA 2004). A discussion of advantages and limitations of the absorbed dose concept and other options for characterizing energy deposition is provided by the ICRU (2011b).

1.1.4 Internal exposure assessment

The first step of internal exposure assessment is to perform measurements of the activity of radionuclides of interest present in the body and their distribution among tissues and organs. These are either taken indirectly from samples of the person (*in vitro*) and its environment (IAEA and ILO 1999; ICRP 1997), or directly on the person (*in vivo*) with *body counting* (IAEA 1996). While *in vitro* methods can detect any type of radiation emitted from the sample, *in vivo* methods are restricted to gamma and X-rays (and high-energy beta rays) due to high attenuation of other types of radiation in the human body. However, *in vivo* measurements with a multi-detector setup have the advantage that they can quantify an activity distribution among several parts of the body, usually lungs, liver, skeleton and thyroid. Additional information about the exposure scenario, for example, incorporation pathway, potentially incorporated radionuclides and their molecular structure, and approximate time and duration of exposure, is collected.

Following the activity assessment, an appropriate biokinetic model is selected and applied to estimate the *intake*, i.e. the total incorporated activity. Based on this value, the biokinetic model is applied again to derive a time-dependent activity distribution for typically 50 years following the incorporation. The integration of these values over time gives a total activity for each organ. This is evaluated with a dosimetric model that relates organ activity to organ absorbed dose depending on type of radiation and anatomy of a reference person. The result is the *effective dose* to that person.

1.1.5 Body counting

Body counting is a form of gamma-ray spectroscopy with a set of energy-sensitive radiation detectors, e.g. scintillation or semiconductor detectors, and a person with an incorporated source in a shielded chamber. The number of interaction events in a region of interest of the recorded pulse-height spectrum of the detector per unit time is quantified as *count rate*.

For *whole body counting*, the detectors are arranged to cover all regions of interest of the body in one measurement. This method is primarily used for routine monitoring of workers, when time-efficient measurements are necessary, for radionuclides that are homogeneously distributed in the human body, such as ^{137}Cs in muscle tissue, or for radionuclides that have high-energy photon emissions, such as ^{60}Co, ^{137}Cs and ^{154}Eu. This is achieved by positioning detectors in a relatively large distance to the person.

Partial body counting, on the other hand, targets only selected structures, for example, thorax for lungs, upper abdomen for liver, neck for thyroid, or knee and head as representatives for skeleton, but provides lower uncertainties and higher spatial resolution because of the small distance between region of interest and detector. This also allows the detection of low-energy photons.

1.1.6 Body counter calibration

Radiation detectors are sensitive to the energy of the emitted particles, the relative location of the source, and the shape of the person because of scattering and attenuation effects in the body. They need to be calibrated to give reasonable results. For this purpose, *physical phantoms* containing defined amounts of certain radionuclides are used. The *counting efficiency* of a detector in a specific measurement setup is calculated as the count rate per unit activity of the radionuclide. The similarity of calibration and measurement scenario is a measure of the uncertainty of the given results.

In body counter calibration, physical phantoms are anthropomorphic and composed of tissue-equivalent materials. Parts can be exchanged for radioactive equivalents. The production of anatomically accurate physical phantoms (ICRU 1992b) is very challenging, but an alternative to physical body counter calibration is offered by computer simulation. *Computational phantoms* are primarily created from computed tomography of medical patients, and radiation detectors and other structures are modelled using solid geometry modelling techniques (Mortenson 1985). Radiation transport simulation offers accuracy (in energy ranges where experimental data is available) and computational efficiency for arbitrary exposure and measurement scenarios.

However, high uncertainties arise when applying calibration factors obtained from reference measurements to real measurement scenarios. The main factors are differences in anatomy of phantom and person, inhomogeneous activity distributions, and changes in detector positions due to difficulties with the exact reproduction of positions and due to different body shapes. These uncertainties are more significant for low-energy photons ($<100\,$keV), because of their increased attenuation in tissue.

1.1.7 Personalisation methods

In the case of radiation accidents with doses approaching regulatory limits, it is necessary to provide dose estimates that are specific to the individual and the circumstances of the exposure. Several methodologies for personalised body counter calibration (Doerfel, Heide and Sohlin 2006; Henriet et al. 2012; Lynch 2011; Mohr and Breustedt 2007; Pierrat et al. 2007) have been proposed using non-invasive medical imaging (e.g. ultrasound or magnetic resonance imaging) to build case-specific models for computational body counter calibration or to quantify properties of anatomic structures with *anthropometric parameters* and correlate these to counting efficiency.

In the case of lung counting, *chest wall thickness* (Sumerling and Quant 1982) has been identified as an anthropometric parameter with high sensitivity to counting efficiency. Samples of body mass and height are used to build statistical models to estimate chest wall thickness and adjust counting efficiency accordingly. These methods have also been incorporated into the development of physical torso phantoms, which can be usually extended with chest overlays of various thicknesses and muscle-adipose tissue ratios to modify chest wall thickness.

1.2 Objectives

The goal of this work is to quantify and reduce uncertainties in activity assessment with partial body counting due to variation in human anatomy to improve dose estimates for individuals exposed to radiation from radionuclide intake.

The basic approach to achieve this goal is the development of a personalisation framework based on body counter calibration with computational phantoms that represent a broad range of variations in human anatomy. The results of sensitivity analyses lead to the creation of models estimating counting efficiency for given anthropometric parameters. Additionally, guidelines for technicians performing body counter calibration and measurement of anthropometric parameters are defined to improve reproducibility of the results and to ensure the applicability of the framework.

1.3 Scope

The body counter of the *in vivo* measurement laboratory (IVM) at Karlsruhe Institute of Technology (KIT) was selected for the application of the framework. The main task of the laboratory is routine monitoring of occupationally exposed individuals, but it is also capable of handling emergency situations and accidents.

The measurement system consists of four high-purity germanium detectors with an operational energy range of about 10 keV to 2 MeV. Two partial-body measurement setups are specified: (1) 2×lungs, liver and knee, and (2) 4×head. Calibration for lung and liver counting is performed with a torso phantom with variable chest wall thickness. The current personalisation method is designed for a legacy system of two phoswich detectors. The ratio of mass and height of a person are measured to estimate its chest wall thickness, which is then used to adjust counting efficiency with respect to the sensitivity observed on the phantom.

In this context, the application of this work focuses on incorporation in lungs, liver and skeleton of adults, and equipment available at the laboratory. The intended use of the framework is to guide technicians in the calibration of detectors in partial-body measurement setups to derive results related to the individual. Especially, in the low-energy range, the framework is supposed to give better estimates than current personalisation methods.

1.4 Outline

The contents of this work is organized into four main parts:

Part I Introduction introduces the topic of *in vivo* activity assessment with partial body counting and methods for uncertainty estimation and reduction using computational body counter calibration and quantification of anatomic properties. Additionally, methods for non-parametric regression analysis and feature selection are described with regard to sensitivity analysis.

Part II Development summarizes and discusses the main focus points of the state of the art in personalisation methods for body counting and describes the approach of this work. This leads to a framework for personalised body counter calibration based on correlation of counting efficiency and anthropometric parameters. The associated methods and materials required for its implementation involve the design of a data model and a software tool to assist in modelling, simulation and evaluation of radiation transport scenarios, and the related procedures for computational body counter calibration, assessment of anthropometric parameters, and data analysis.

Part III Application describes the application of the framework to the calibration of the IVM body counter with different types of computational phantoms. Body counter and phantoms are modelled and processed with focus on accuracy, reproducibility and computational efficiency. The resulting data is used as samples to create statistical models correlating counting efficiency and anthropometric parameters, analyse different sources of uncertainty, and perform measurement validation.

Part IV Discussion discusses and concludes the application of the developed framework, and sets the results into context regarding their transferability to actual measurements. Also, the application of the developed data model and software implementation is discussed, and possible improvements and future developments are described.

2 Radioactivity

Since photons are of major interest for body counting, it is important to understand how they are produced by radioactive decay and how they interact with matter. The goal of this chapter is to summarize the main interaction processes that comprise photon transport with regard to measurements with gamma-ray spectroscopy and to define dosimetric and operational quantities for future reference. Knowledge about these aspects also gives context to the principles of radiation transport codes using the Monte Carlo method.

Interaction effects of photons with matter cause scattering, absorption and the generation of secondary particles. For a narrow photon beam, this results in broadening and attenuation. The probabilities of these interactions depend on thickness, density, and effective atomic number (i.e. chemical composition) of the matter, and energy of the photons. Similar to radioactive decay, radiation transport is a stochastic process which behaves deterministically for a large number of particles.

2.1 Radioactive decay

Radioactive decay is the process of spontaneous transformation of an atom with *unstable* nucleus under emission of energy in form of radiation. Unstable atoms are called *radionuclides*. The product of the decay can be the same radionuclide in a different state, or a different isotope or chemical element. The basic decay modes of radionuclides are *alpha*, *beta*, and *gamma* decay. Detailed descriptions of radioactive decay processes are given in standard literature (Attix 1991; Knoll 2010; Reilly et al. 1991).

2.1.1 Decay modes

The decay of radionuclides is primarily defined by the emitted particle and the according change of the nuclear structure:

Alpha decay is accompanied by the emission of an α particle, which is a positively charged helium ion $^4_2$He. The radionuclide decays into an atom with mass number decreased by 4 and atomic number decreased by two. The typical kinetic energy of alpha particles is about 5 MeV.

Beta decay is accompanied by the emission of a β particle, which is either an electron e^- or positron e^+, and an electron anti-neutrino $\overline{\nu}_e$ or electron neutrino ν_e, respectively. Accordingly, this increases or decreases the atomic number of the atom by one. The emitted beta particles have a continuous kinetic energy spectrum.

Gamma decay is accompanied by the emission of a γ quantum, which is a photon. This process is usually preceded by alpha or beta decay, which leaves the nucleus in an excited nuclear state.

2.1.2 Activity

From observations it is known that in a large sample of N atoms of a radionuclide the number of decay events $-\mathrm{d}N$ in a small time interval $\mathrm{d}t$ is proportional to the number of atoms (ICRU 2011a). This quantity is called *activity* A. The unit of activity is s^{-1} with the special name Becquerel (Bq). The proportionality constant is the *decay rate* λ given per unit time.

$$A = -\frac{\mathrm{d}N}{\mathrm{d}t} = \lambda N \tag{2.1}$$

This deterministic model approximates the random decay process for large sample sizes. N is the initial number of atoms at time $t = 0$.

$$N(t) = N\mathrm{e}^{-\lambda t} \tag{2.2}$$

2.1.3 Decay data

Nuclear structure and decay data for radionuclides is usually stored in Evaluated Nuclear Structure Data File (ENSDF) format (Tuli 2001), which is maintained by the National Nuclear Data Center (NNDC) at Brookhaven National Laboratory (BNL). Several databases (Laboratoire National Henri Becquerel 2013; National Nuclear Data Center 2013a) are available that provide data in ENSDF format compiled from recent publications of experimental results or calculations.

2.2 Radiation transport

When ionizing radiation passes through matter, particles have discrete interaction effects depending on particle type and energy. Interactions basically transfer energy from ionizing particles to atoms of the medium. The differences in these particles are mainly charge (positive, negative, or neutral), and mass or size. Heavy, slow, or charged particles generally have larger interaction probabilities than light, fast, or uncharged particles.

Charged particles (e.g. electrons, protons and alpha particles) directly interact with atomic nuclei, orbital electrons, or other charged particles through their electrical field. They are called *directly ionizing* radiation. Uncharged particles (e.g. photons and neutrons) only interact via collisions. Elastic collisions cause an energy transfer between the two particles and scattering. Inelastic collisions result in absorption of one particle and acceleration of the other. They can also indirectly cause ionization via secondary charged particles produced by interaction effects. They are called *indirectly ionizing* radiation.

Most particle interactions transfer energy to nuclei or their orbital electrons resulting in *excitation* or *ionization*. Excitation refers to an increased energy state of a particle with respect to its ground state. The spontaneous relaxation of the particle to a lower energy state is usually accompanied by emission of photons releasing the excess energy. An orbital electron can also be ejected from the electron cloud of an atom causing its ionization. Excitation and ionization of electrons create electron vacancies or *electron holes*. These holes are filled by higher orbital electrons, also causing emission of photons.

2.2.1 Absorbed dose

The difference in energy between ionizing particles entering and leaving a single interaction is the *energy deposit*. The *energy imparted* to the matter in a given volume is the sum of all energy deposits in that volume. *Absorbed dose* is the mean energy imparted in a certain quantity of medium by radiation per unit mass. This is a basic quantity in radiation protection used to quantify radiation exposure and relate it to biological damage and health risk to people. Detailed definitions of these and derived quantities are provided by ICRU (2011a).

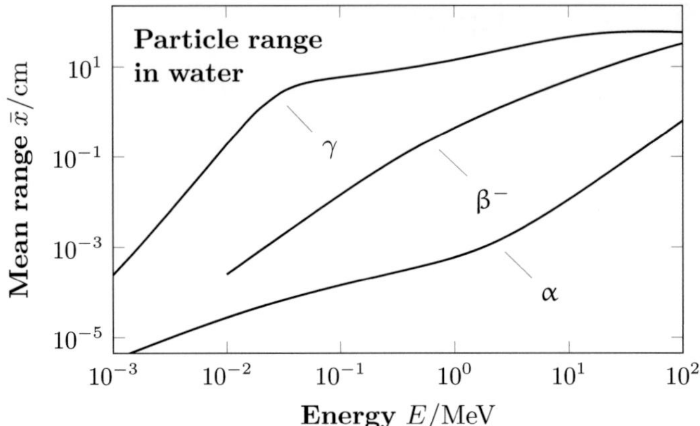

Figure 2.1: Mean range of particles in water as a function of kinetic energy. For α and β^- particles the continuous-slowing-down approximation (CSDA) is used based on data from Berger et al. (2005), and for γ particles the inverse linear attenuation coefficient is used to approximate the mean free path based on data from Berger et al. (2010).

Radiation transport with regard to the focus of this work mainly takes place in the human body. The range of α and β^- particles with typical respective energies of about 5 MeV and 300 keV (ICRP 2010) in water, which is comparable to adipose, muscle, and general soft tissue, is less than 1 mm (figure 2.1). However, these particles undergo many scattering events before being fully absorbed. These interactions lead to excitation and ionization of atoms which produce secondary particle emissions. Only photons (or high-energy electrons) are able to leave the body to be detected by gamma-ray spectroscopy.

2.2.2 Energy fluence

Radiation fields are generally characterized by type, energy and intensity of its particles. An important quantity for measuring the intensity of a radiation beam is *fluence*. The fluence Φ is the number of particles $\mathrm{d}N$ incident on a sphere of cross-sectional area $\mathrm{d}A$ (ICRU 2011a). The unit of fluence is m^{-2}.

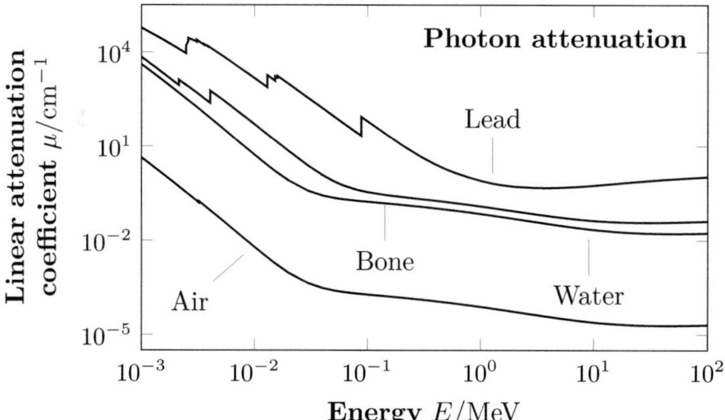

Figure 2.2: Linear attenuation coefficients for photons in different media as a function of energy based on data from Berger et al. (2010). The attenuation for adipose, muscle, and general soft tissue is very similar to water. The attenuation for bone is significantly higher.

$$\Phi = \frac{\mathrm{d}N}{\mathrm{d}A} \tag{2.3}$$

2.2.3 Linear attenuation coefficient

The *linear attenuation coefficient* μ describes the mean attenuation per unit length of a *narrow* radiation beam with fluence Φ_0 incident on a medium (figure 2.2). The fluence $\Phi(x)$ of the attenuated beam at distance x into the medium along the initial beam direction is exponentially decreasing with increasing effective thickness μx.

$$\Phi(x) = \Phi_0\,\mathrm{e}^{-\mu x} \tag{2.4}$$

Linear attenuation coefficients can be experimentally measured for chemical elements and are available in form of tabulated *mass attenuation coefficients* μ/ρ (Berger et al. 2010) normalized to density ρ.

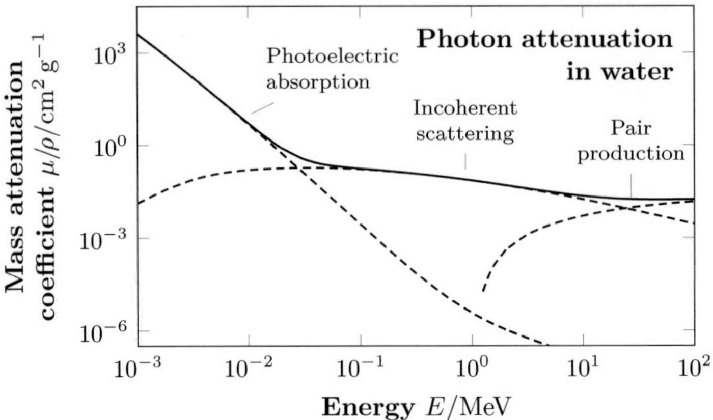

Figure 2.3: Contribution of different interaction effects of photons in water to the total mass attenuation coefficient as a function of energy based on data from Berger et al. (2010).

Considering a mixture or compound medium with weight fractions $w_i > 0$ of chemical elements i with mass attenuation coefficients $(\mu/\rho)_i$ and $\sum_i w_i = 1$, the total mass attenuation coefficient μ/ρ of the medium is the weighted sum of the fractions. This is an approximation which is sufficient for media with low effective atomic number and low photon energies (ICRU 2008).

$$\mu/\rho = \sum_i w_i \, (\mu/\rho)_i \tag{2.5}$$

2.2.4 Cross section

Given a particle with incident particles of fluence Φ that produce a number N of interaction events, the cross section σ is a measure for the probability of an interaction event to occur (ICRU 2011a). It can be interpreted as the effective area around the particle that would lead to an interaction when an incident particle crosses it. The unit of cross section is m^2. The cross section is proportional to the mass attenuation coefficient μ/ρ.

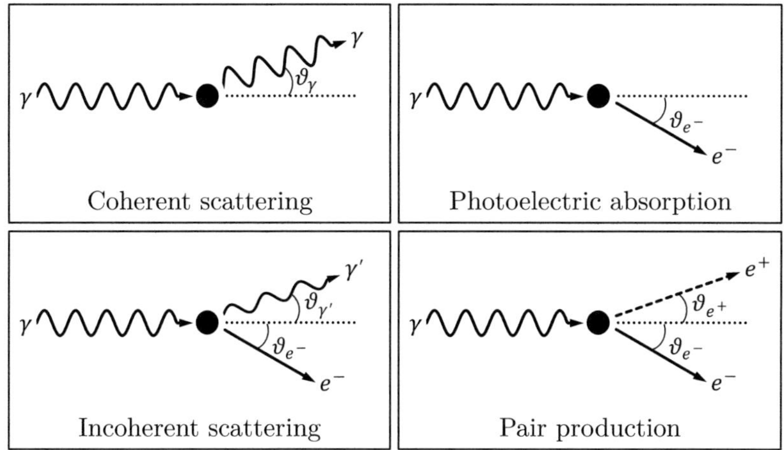

Figure 2.4: Schematics of interaction effects of photons with matter based on Salvat, Fernández-Varea and Sempau (2011).

$$\sigma = \frac{N}{\Phi} \propto \mu/\rho \tag{2.6}$$

The total photon cross-section σ_γ is the combination of the cross sections σ_i of the individual interaction effects i of photons with matter (figure 2.3). These are *photoelectric* (and *photonuclear*) absorption, *coherent* and *incoherent* scattering, and *pair* (and *triplet*) production. Detailed descriptions of these effects are given by Attix (1991), Reilly et al. (1991), Kawrakow et al. (2011), and Salvat, Fernández-Varea and Sempau (2011).

$$\sigma_\gamma = \sum_i \sigma_i \tag{2.7}$$

In the following, short descriptions of the dominant interaction effects in the energy range relevant for body counting (about 10 keV to 2 MeV) are given. These are photoelectric absorption, incoherent scattering, and pair production (figure 2.4).

2.2.5 Photoelectric absorption

Photoelectric absorption occurs when an incident photon undergoes an inelastic collision with an orbital electron, causing its ejection from the atom. The photon is fully absorbed in the process. This means that the energy E_γ of the photon has to be large enough to raise the electron to a positive energy state which is at least the binding energy E_{e^-} of the electron. The excess energy $\Delta E = E_\gamma - E_{e^-} > 0$ is converted to kinetic energy of the electron. At low energies close to the binding energies of the electrons, there are abrupt changes in interaction probabilities when the excess energy reaches the binding energy of the next orbital. These changes are visible as *absorption edges* in the cross sections (figure 2.2).

2.2.6 Incoherent scattering

Incoherent or *Compton* scattering occurs when an incident photon undergoes an inelastic collision with an orbital electron, causing its ejection from the atom. A portion of the energy ΔE of the photon is transferred to the electron in the process. The energy in excess to the binding energy is converted to kinetic energy of the electron. The photon is scattered at angles of up to 180°. The amount of transferred energy depends on the scattering angle ϑ with $\Delta E \propto 1/\left(1 - \cos\vartheta\right)$.

2.2.7 Pair production

Pair production occurs when an incident photon interacts with the electric field of a nucleus and transforms into an electron-positron pair. The photon is fully absorbed in the process. This means that the energy E_γ of the photon has to be larger than the equivalent energy of the rest masses $E_e = E_{e^-} = E_{e^+}$ of electron and positron. The exceeding energy $\Delta E = E_\gamma - 2E_e > 0$ is shared by both particles as kinetic energy. With $E_e \approx 0.511\,\mathrm{MeV}$, it follows that $E_\gamma > 1.022\,\mathrm{MeV}$ for this process to occur.

2.3 Photon production

Photons are mainly produced through *spontaneous emission*. Other effects that generate photons are *bremsstrahlung* and *electron-positron annihilation*. A detailed overview of particle production processes is given by Attix (1991).

2.3.1 Spontaneous emission

Spontaneous emission is the transition of a nucleus or an orbital electron in an excited energy state E_2 to a lower energy state E_1 with $E_2 > E_1$ by releasing excess energy $\Delta E = E_2 - E_1$ in form of a photon with energy ΔE. Photons emitted by excited nuclei are called *gamma rays*, and photons emitted by excited electrons are called *X-rays*.

In the case of X-rays, the energies of the emitted photons are characteristic because of the *discrete* energy states of orbitals given by the Rutherford-Bohr model as $E_n \propto -Z^2/n^2$ with atomic number Z and principal quantum number n.

The energies of gamma rays are determined by alpha or beta decays preceding the gamma decay, which leave the daughter nuclide in an excited state. This usually leads to much higher energies than those of X-rays.

2.3.2 Bremsstrahlung

When a charged particle moves relative to an electric field (e.g. in vicinity of a nucleus or an electron), it interacts with that field by transferring energy through attraction or repulsion. The particle is scattered and the transferred energy is released by the electric field in form of a photon, called bremsstrahlung.

Bremsstrahlung occurs in X-ray tubes where electrons are accelerated by an electric field and shot into a metal target. The electrons eject inner orbital electrons of the metal atoms which release characteristic X-rays upon relaxation. In addition, the electrons are decelerated in the target through interaction with the electric field of the nuclei and emit a *continuous* spectrum of bremsstrahlung.

2.3.3 Electron-positron annihilation

When a positron collides with an electron, both particles are annihilated, mostly releasing their energy in form of two photons. Of course, this process must conserve electric charge, energy, and linear and angular momentum. Therefore, the energy E_γ of the photons is determined by the rest energy E_e of electron and positron. In the basic case, both photons have an energy of $E_\gamma \approx 511\,\text{keV}$ and move in opposite directions. For the annihilation of particles with higher kinetic energies also other particle pairs or single particles can be produced, given sufficient energy is available.

3 Gamma-ray spectroscopy

Gamma-ray spectroscopy is the detection and quantization of radionuclide sources emitting gamma rays. Understanding the principles of photon detection with semiconductor detectors, which are mainly used today, and spectral analysis gives insight into body counting, which is necessary for modelling detection systems for radiation transport simulation. Knowledge about calibration procedures, which include energy and energy resolution calibration, is necessary to produce accurate and reproducible measurement results.

3.1 Semiconductor detectors

Semiconductor detectors are basically diodes with a p-i-n junction (figure 3.1). The *intrinsic* (i) region consists of the basic semiconductor material. Diffusing or implanting certain materials creates a deficiency (p+) or an excess (n+) of electrons in the *valence band* of these regions.

Applying a *reverse bias* v_e, i.e. a positive voltage to the n+ region and a negative voltage to the p+ region attracts electric charges to the respective electrodes, which increases the width of the intrinsic (or *depletion*) region, and the junction becomes an insulator (Canberra 2008).

Ionizing radiation incident on the depletion region ionizes electrons in the valence band, which move to the *conduction band* leaving an electron hole. The minimum energy required to free a valence electron is equal to the band gap. All or part of the remaining photon energy is converted into kinetic energy of the electron. The ionized electron interacts with other valence electrons and produces additional ionizations (Reilly et al. 1991). Because of the electric field, the electrons move to the n+ electrode and the electron holes move to the p+ electrode. The induced current creates a voltage pulse.

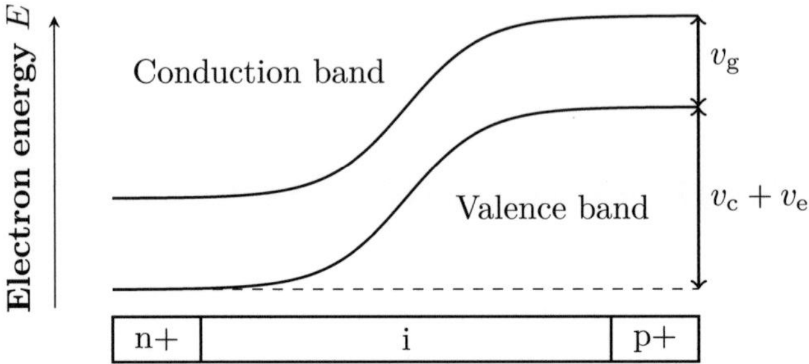

Figure 3.1: Structure of the p-i-n junction of a semiconductor detector. The p+ and n+ regions act as electrodes for charges released by interactions of ionizing radiation in the intrinsic region (i). v_c is the (negative) leakage potential of the diode, v_e is the (positive) externally applied potential, and v_g is the band gap potential. Graphic based on Knoll (2010).

3.1.1 Pulse height

Counting the number of released charges by a single photon with a charge sensitive preamplifier gives the energy of the photon as a multiple of the band gap. The number of charges released by a single photon, which is counted with a charge-sensitive preamplifier, is proportional to the energy of the incident photon. This may lead to pile-up effects for high activities. However, these are usually not encountered in body counting.

3.1.2 Dead layer

The electrodes form an inactive region or *dead layer* in comparison to the active depletion region, because charge generation by ionizing radiation is not possible in these regions. Photons have to pass the entrance window and the portion of the n+ contact at the front of the crystal to reach the active volume (figure 3.2) and generate a detectable pulse, which is improbable for low-energy photons. A thin dead layer increases the sensitivity of the detector in this energy range.

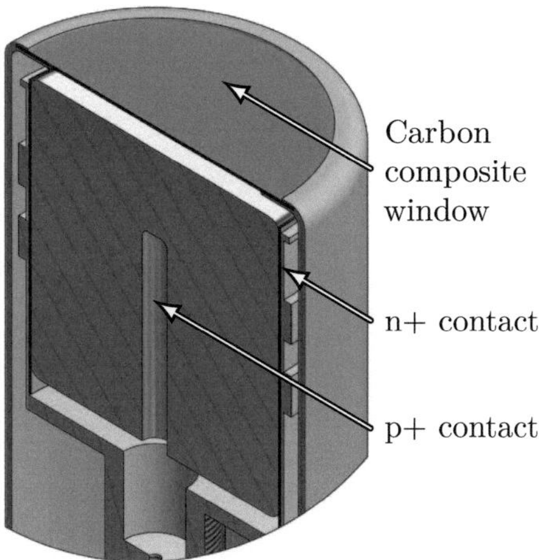

Figure 3.2: Assembly of an XtRa-type detector (Canberra 2013). The p+ electrode contact is at the hollow inside of the crystal, and the n+ electrode contact is at the outer surface of the crystal. The part of the contact at the front of the crystal has been mechanically removed.

3.1.3 Cooling

Compared to silicon as a base material, germanium has a higher density and crystals are produced with high purity. The depletion region is large, allowing only a thin dead layer. The band gap of germanium is very low at room temperature with 0.67 eV. This allows many valence electrons to move to the conduction band simply due to their thermal energy, which creates an inverse leakage v_c and results in a low signal-noise ratio in the pulse-height spectra. To compensate this effect, germanium detectors are cooled down to 77 K during operation using liquid nitrogen or thermoelectric cooling (Canberra 2012), which increases the band gap to 2.96 eV.

3.2 Detector calibration

The dominating interaction effects of photons with germanium in the depletion region and in the energy range of up to 2 MeV are photoelectric absorption and incoherent scattering. The detector response to a mono-energetic point source does not contain only one discrete line, but follows a characteristic distribution which basically consists of a photo peak and a Compton background continuum.

If the incident photon receives photoelectric absorption, all of its energy in excess of the band gap is converted to kinetic energy of the valence electron. If, on the other hand, the photon is scattered, only a portion of its energy is transferred with the maximum at a scattering angle of 180°. Photons may be scattered multiple times before either being absorbed or leaving the depletion region. Other effects are energy loss due to heating of the lattice crystal structure and thermal noise in the charge integration. This broadens the photo peak. A full description of the features of a pulse-height spectrum is provided by several textbooks, such as Reilly et al. (1991) or Knoll (2010).

3.2.1 Energy resolution

Given the pulse-height spectrum of a detector in response to a mono-energetic point source with energy E_0, the measured energy E of the detector is modelled as a random variable with normal distribution $N(\mu, \sigma^2)$ with mean $\mu = E_0$ and standard deviation $\sigma = E_{\mathrm{FWHM}}/(2\sqrt{2 \ln 2})$. FWHM refers to the *full-width-at-half-maximum* of the photo peak (figure 3.5).

A non-linear model (Reilly et al. 1991) is usually used to describe the energy dependence of E_{FWHM}. Several samples for energy E and E_{FWHM} are directly measured for the photo peaks of known mono-energetic point sources, and are used to characterize the detector with parameters a, b, and c (figure 3.3).

$$E_{\mathrm{FWHM}} = a + b \sqrt{E + c E^2} \tag{3.1}$$

3.2.2 Energy channels

The domain of the amplifier voltage pulse of a detector is discretized for feasibility of measurement analysis into a certain number of equidistant *channels* appropriate to the resolution of the detector. Each detection event is categorized in exactly one of those channels. The voltage induced by incident radiation is generally proportional to the energy deposited, but

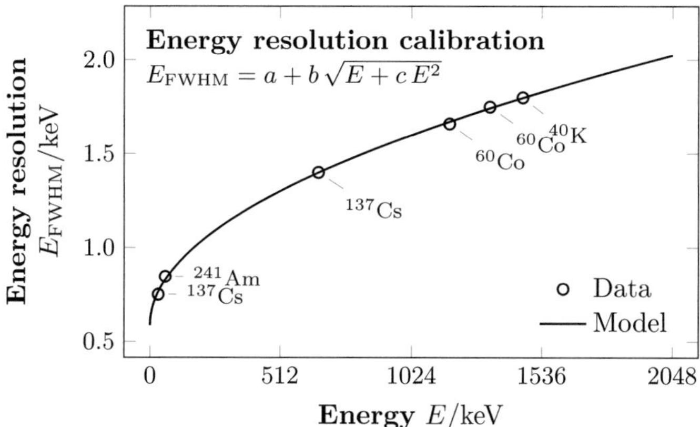

Figure 3.3: Energy resolution calibration of an HPGe detector at IVM with sample measurements. Samples of corresponding energy and FWHM are determined by analysing the full energy peaks of measured point sources and fit to an according model.

calibration is needed to give exact values. Each energy E of a detection event is assigned a channel number C with a polynomial of order two with parameters a, b, and c.

$$E = a + b\,C + c\,C^2 \tag{3.2}$$

The non-linear contribution, however, is comparably small. Radionuclide point sources with known photon energies are used to provide samples of channel and energy (figure 3.4).

3.3 Measurement evaluation

The pulse-height spectrum of a detector is given as the number of *counts* N_i for a given channel $i \in \{1, 2, \dots\}$ with energies $(E_{i-1}, E_i]$. The unit of counts is dimensionless, but usually given as counts.

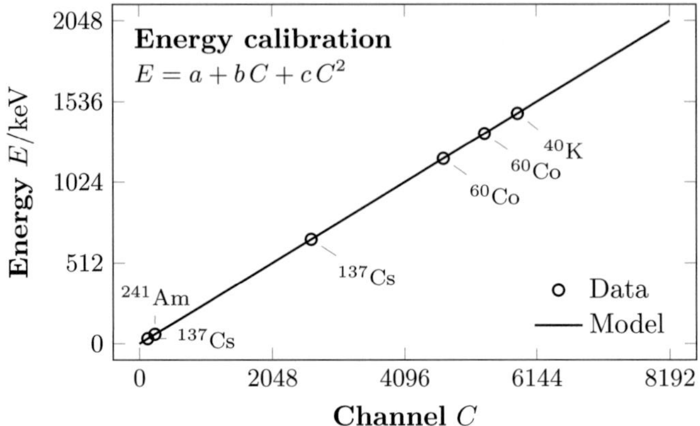

Figure 3.4: Energy calibration of an HPGe detector at IVM with sample measurements. Samples of corresponding channel and energy are determined by analysing the full energy peaks of measured point sources and fit to an according model. The channel width is 0.25 keV in this case.

3.3.1 Counting statistics

Since radioactive decay is a stochastic process, the *count rate* $cps = N/\Delta t$ induced in a detector at energy E by the activity A of a source with measurement time Δt is the sum of identically distributed independent random variables modelling the potential for individual photons being detected. According to the central limit theorem (Koroliuk 2013), the count rate is a random variable with normal distribution $N(\mu, \sigma^2)$ whose mean μ is unbiased and whose variance $\sigma^2 \propto 1/N$ converges with increasing number of detection events (i.e. an increase in measurement time) to zero. This also holds for an energy channel, whose number of counts is the sum of all counts of energies in that channel.

3.3.2 Peak analysis

The first step of spectral analysis is the identification of photo peaks (Reilly et al. 1991) resulting in the peak centroids. The next step is peak analysis. This is the process of estimating the number of counts of a peak considering energy resolution, counting statistics, and Compton

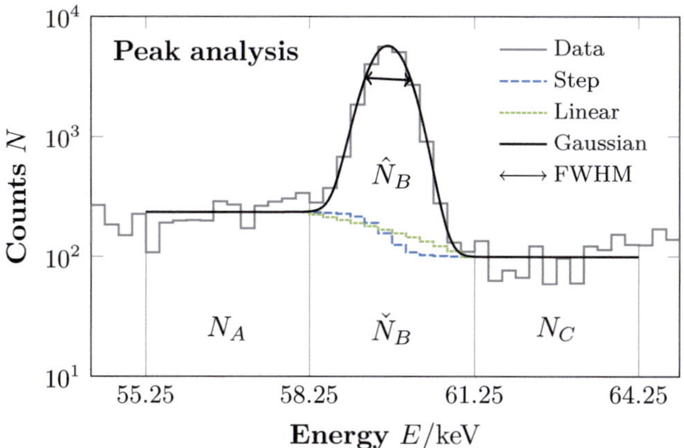

Figure 3.5: An example spectrum of the main gamma peak of ^{241}Am of an HPGe detector at IVM and two models for background approximation: linear and step function. The Gaussian fit is annotated with its FWHM (sloped due to the asymmetric background). Regions of interest are shown for the peak area (B), and left (A) and right (C) background.

background from other peaks. The general concept is to define regions of interest (figure 3.5) for the peak depending on FWHM at that energy. Additional regions are defined for estimation of the background in the direction of higher and lower energies. Then, the detection background is approximated with an appropriate model, and the spectrum is corrected by subtracting the background. The net counts of the peak are calculated by summing the corrected spectrum in the peak region. Common models for approximating detection background are constant, linear, polynomial, and step models (International Organization for Standardization 2010a).

3.3.3 Background subtraction

Given a spectrum N of a detector characterized with energy resolution E_{FWHM} and a single peak at E_0. The net counts $\hat{N}(E_0)$ of the peak are estimated by background subtraction (Canberra 2006):

1. Define adjacent regions of interest A, B, and C comprising multiple channels as shown in figure 3.5. A suggested default value for all regions is $\Delta E = 2.5\,E_{\mathrm{FWHM}} \approx 5.875\,\sigma$.

2. Align those regions to the channels by rounding to multiples of the channel width.

3. Determine the number of counts $N_X = \sum_{i \in X} N_i$ for each region X.

4. Choose a model representing the background $\check{N}_i$ for each channel i in region $B = \{i_0 + 1, \ldots, i_0 + b\}$. Examples are linear f_{linear} or step f_{step} continua.

5. The number of counts in the peak region with subtracted background is $\hat{N}(E_0) = \hat{N}_B$.

$$\hat{N}_B = N_B - \check{N}_B \tag{3.3}$$

$$\check{N}_i = \frac{N_C}{|C|} + f(i - i_0)\left(\frac{N_C}{|C|} - \frac{N_A}{|A|} \right) \tag{3.4}$$

$$f_{\mathrm{linear}}(i) = \frac{i}{|B| + 1} \tag{3.5a}$$

$$f_{\mathrm{step}}(i) = \frac{\sum_{j=1}^{i} N_j}{N_B} \tag{3.5b}$$

Various methods for radionuclide identification, peak analysis, and computation of detection limits are specified in standards (International Organization for Standardization 2010b; International Organization for Standardization 2010a), and provided by dedicated tools, for example, the spectrometry software GENIE 2000 (Canberra 2009).

4 Classical body counting

Body counting is a method for measuring the amount of radioactivity within the human body using gamma-ray spectroscopy. Several detectors are placed relative to the body, targeting single or multiple structures. This allows the quantification of the activity of incorporated radionuclides with spatial resolution in contrast to *in vitro* methods. The term *classical body counting* or simply body counting refers here to experimental measurements, while *computational body counting* refers to computer simulation of the counting procedure.

The main drawback of body counting is the large attenuation of radiation emitted inside the body. Combined with short measurement times to reduce discomfort for the person, a natural background of ^{40}K in muscle tissue, and usually low incorporated activities, this results in high uncertainties in the measured spectra. In partial-body setups, detectors are therefore placed as close as possible to the target structure to improve counting statistics by minimizing attenuation in surrounding air. This amplifies other contributions to uncertainty which may not be apparent in whole-body setups: variation in body size and shape, and non-uniform distributions of radionuclides in the target structure.

4.1 Body counter calibration

In addition to the individual energy and energy resolution calibration, each detector is calibrated in the specific measurement setup to relate the observed peak count rates to source activities. This involves a measurement to determine the detection background, and a measurement to determine *counting efficiency*. Both calibration measurements must be similar to the real measurement in terms of geometry, because the difference between calibration and application basically determines the uncertainty of the estimated incorporated activity.

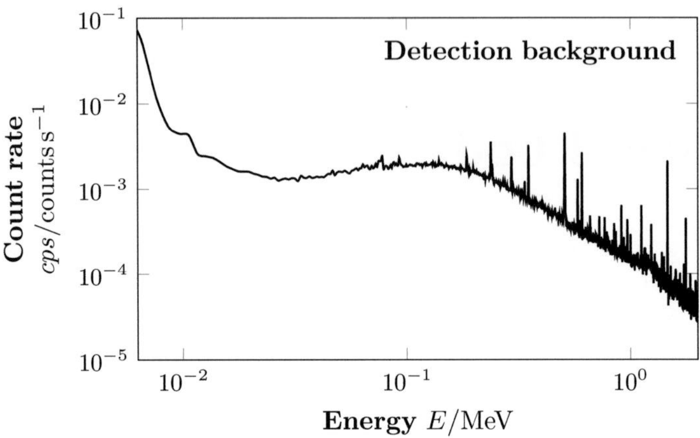

Figure 4.1: Typical background spectrum measured for an HPGe detector at IVM with an empty measurement chamber. A pile-up of overlapping Compton backgrounds is apparent for low energies. Isolated peaks indicate decay of naturally occurring radionuclides in the chamber.

4.1.1 Background calibration

An issue for detector measurements is the permanent background due to cosmic and terrestrial radiation. To calibrate a detector, the background spectrum (figure 4.1) is measured regularly and then subtracted from non-background measurements. They are usually performed with an empty measurement chamber or with an inactive phantom with a long measurement time to achieve low uncertainties.

From an engineering point of view, high background noise can be avoided by selecting an appropriate location (e.g. an underground room), shielding, and low-background materials for the detector system. Additionally, air filtration and circulation is useful to reduce the concentration of ^{222}Rn and its progeny naturally occurring in air. A detailed overview of background reduction techniques is given by ICRU (2003).

4.1.2 Counting efficiency

Counting efficiency $\eta_{t \leftarrow s}$ for a measurement setup $t \leftarrow s$ with detector t and source s is defined as the number of detection events N_t relative to the number of decay events N_s in the period Δt. This is equivalent to the ratio of corresponding count rate $cps_t = N_t/\Delta t$ and activity $A_s = N_s/\Delta t$. The unit of counting efficiency is dimensionless, but is usually given as counts $\text{s}^{-1}\,\text{Bq}^{-1}$, or counts decay^{-1} or counts photon^{-1} to differentiate between normalization to the general number of decay events or only the number of decay events that produce photons. Counting efficiency can also be interpreted as the probability of a decay event being detected.

$$\eta_{t \leftarrow s} = \frac{cps_t}{A_s} \tag{4.1}$$

Considering the measurement of a radionuclide source with activity A_s and measurement time Δt, $\hat{N}_t$ is the background-corrected number of counts in a photo peak of the radionuclide at energy E with yield y_E, and $\widehat{cps}_t$ is the corresponding count rate. $\hat{N}_s$ is the number of photons produced by s with energy E, and $\hat{A}_s$ is the corresponding activity.

$$\hat{A}_s = y_E\, A_s \tag{4.2a}$$
$$\widehat{cps}_t = y_E\, cps_t \tag{4.2b}$$

For efficiency calibration, source activity must be known and counting efficiency is determined using the count rates at selected photo peaks with high yield. After that, activity assessment can be performed.

4.1.3 Counting efficiency curve

Counting efficiency depends on the structure of the detector. Various detector types have been developed for specific applications (Canberra 2008). It also depends largely on the relative source location, the geometry of inactive scatter material between source and detector, and the energy of the emitted photons.

For a photon to be detected, it must be transported to the active volume of the detector crystal, and then be (at least partially) absorbed. Kramer (2007) describes this with the following analogy: Given a photon beam that is attenuated by two absorbers where the first absorber represents scattering material and the second absorber represents the active volume of the detector. Let Φ_0 be the initial fluence of the photon beam, and Φ_1 and Φ_2 the fluences after penetration of the first and second absorber respectively. The counting efficiency η of this system is equivalent to the product of the transmission through the first layer Φ_1/Φ_0 and the attenuation in the second layer $1 - \Phi_2/\Phi_1$.

$$\eta = \frac{\Phi_1}{\Phi_0}\left(1 - \frac{\Phi_2}{\Phi_1}\right) \tag{4.3}$$

Using, for example, muscle tissue for the first and germanium for the second layer, and according mass attenuation coefficients, densities, and varying thicknesses for the absorbers, gives the characteristic counting efficiency curve. Several parametric models have been suggested to fit this curve. An overview of these and performance comparisons are given by Gray and Ahmad (1985) and Kramer (2007).

4.1.4 Multi-detector systems

In partial-body setups, each detector targets a single structure whose activity should be determined. The pulse-height spectrum of each detector consists primarily of events from the targeted structure, but also contains unintended contributions or *crosstalk* from structures that are not targeted. For example, detectors in front of the chest targeting the lungs will also pick up events due to incorporation in muscle, skeleton, or liver. A better estimate for the activity in the target structure can be determined by including these contributions in the calculation.

The result of an efficiency calibration for a certain detector setup with detectors T and phantom with source structures S is a matrix $\eta_{T \leftarrow S}$. This matrix specifies counting efficiencies $\eta_{t \leftarrow s}$ from any considered source structure $s \in S$ to any detector $t \in T$. Given a source activity vector A_S, the detector count rate vector cps_T is directly dependent according to equation 4.1.

$$cps_T = \eta_{T \leftarrow S}\, A_S \tag{4.4}$$

When given a measurement cps_T and calibration $\eta_{T \leftarrow S}$, the equation can be solved for A_S to determine the activity distribution. This is only reasonable if the system is either determined $|T| = |S|$ or overdetermined $|T| > |S|$ (i.e. targeting each source structure with at least one detector).

An underdetermined system $|T| < |S|$ results in a vector space of linear combinations of emission distributions, which cannot be evaluated.

4.2 Physical phantoms

Phantoms are models of the human body or anatomical structures used widely in medicine (e.g. radiotherapy, nuclear medicine, and medical imaging), radiobiology, and radiation protection. All of these applications have different requirements, and the functionalities of the phantoms vary largely. In this work, the term *phantom* is restricted to calibration phantoms with internal radiation sources. The term *physical phantom* is used to refer to material phantoms in contrast to *computational phantoms*, which are used in computer simulations.

Especially for low-energy photons, body counting is sensitive to the anatomy of the measured person and must be calibrated accordingly. This is done using physical phantoms, which have a known distribution of a radioactive material with specified activity. The distribution is typically homogeneous in the active parts of the phantom as this is the most general assumption.

Common phantom types in use are brick, bottle, and anthropomorphic phantoms (ICRU 1992b). While brick and bottle phantoms offer a higher flexibility in terms of modifications, anthropomorphic phantoms are by definition anatomically accurate to a certain detail. Phantoms consist of artificial, tissue-equivalent materials with emphasis on density and regard to chemical composition. There are also phantoms with embedded authentic bone from donations of deceased people with incorporations. A detailed overview of physical phantoms for different applications is given by ICRU (1992b).

4.2.1 Anthropomorphic phantoms

The production process of anthropomorphic phantoms is very complex and has high demands on the utilized materials (ICRU 1992b; Traub 2008). The individual structures of the product must be homogeneous (especially the active parts), mechanically and chemically stable, and exhibit no degradation due to routine handling. Typical types are head, neck, torso, and knee

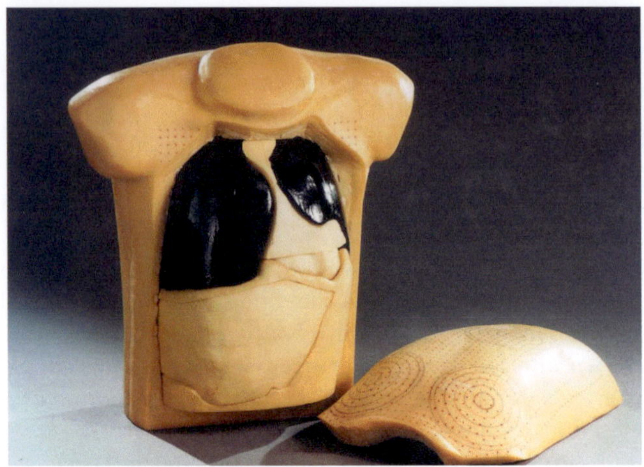

Figure 4.2: A physical torso phantom with a set of chest overlays (Doerfel, Heide and Sohlin 2006). The predefined detector positions for body counter calibration are marked on the chest cover as concentric circles at the left lung, right lung, and liver.

phantoms as these are the accessible structures that contain bone, thyroid, lungs, and liver, and are likely to accumulate certain radionuclides.

Especially, torso phantoms (figure 4.2) are very elaborate compared to other types. They are assemblies of individual parts which can be exchanged for their geometrical equivalents loaded with radionuclides (e.g. natural U, ^{238}Pu, ^{232}Th, or ^{241}Am). This allows the calibration for several energies and also for crosstalk. In addition, torso phantoms have sets of chest overlays with varying thicknesses and different ratios of adipose and muscle tissue to adjust to different groups of individuals.

4.3 Detector positioning

Gamma-ray spectroscopy is very sensitive to the location of the detector relative to the source. This is a problem in body counting, where the source is shielded by thick layers of muscle and adipose tissue, and the detectors may cover multiple small structures (because of their size). When dealing with low count rates, the detector should be placed as close as possible to the source or in light contact with the entrance window to achieve optimal detection limits. Moreover, repeatable and reproducible detector positioning is crucial for reconstruction and comparison of measurements.

ICRU (2003) summarizes basic measurement setups placing the person in a stretcher or chair and targeting chest for left and right lung, upper body for lymph nodes, abdomen for liver, head, knee, or wrist for skeleton, and whole body for muscle tissue. Body counters have predefined measurement setups for several expected radionuclides customized to the specifics of the laboratory. Detectors are usually mounted in arrays and have limited possibilities for adjustments. In some cases, however, the detectors are mounted on individual racks which provide more degrees of freedom.

4.3.1 Skeleton

A major deposition site for uranium and transuranic radionuclides that enter the circulatory system, for example, ^{235}U, ^{238}U, ^{239}Pu, and ^{241}Am, is the skeleton. The common decay product ^{210}Pb can also be measured there. The head is the preferred structure for skeletal measurements, since it has only a thin layer of subcutaneous and muscle tissue, is easily accessible and relatively far from other incorporation sites. Lynch (2011) positions a detector in front of the forehead. An alternative is the measurement of wrists, knees, and ankles due to the higher deposition in trabecular bone compared to cortical bone.

4.3.2 Liver

Another major deposition site is the liver. Lynch (2011) aligns the detectors along the seventh right rib in front of the upper abdomen.

4.3.3 Lungs

The retention time of radionuclides in the respiratory tract, primarily the lungs, is relatively short compared to skeleton and liver. However, lung counting is important for the detection of radionuclides shortly after inhalation. Several positioning strategies are described in literature:

- Sumerling and Quant (1982) suggest detector positions on the left and right side of the sternum centred on the third ribs. The detectors

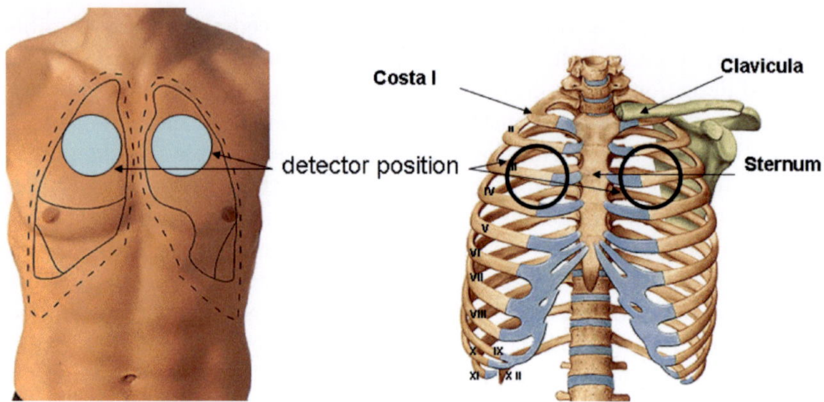

Figure 4.3: Detector positions for lung counting with two germanium detectors (Hegenbart and Gün 2010).

cover the first and second intercostal space, which are easier to penetrate for low-energy photons compared to bone.

- Doerfel, Heide and Sohlin (2006) define positions in detail for a 2×2 array of germanium detectors. The detectors are centred on the centroids of the lungs.

- Farah, Broggio and Franck (2010) position a 2×2 array of germanium detectors. The distance between the left and right pair is fixed to 2 cm and to 1 cm between the upper and lower detectors of a given pair. The inclination of the array is fixed to 35° to bring the detectors as close as possible to the skin. Anatomic landmarks are used to centre the detectors on the lungs in relation to the clavicles depending on the chest girth.

- Hegenbart, Gün and Zankl (2010) position two independent germanium detectors with a diameter of 7.5 cm (figure 4.3). Both detectors are inclined to 25° in order to be parallel to the skin surface. The detector front axes are centred on the third rib, 7 cm away from the centre of the sternum with a skin distance of about 1–2 mm. The setup is symmetrical on both sides.

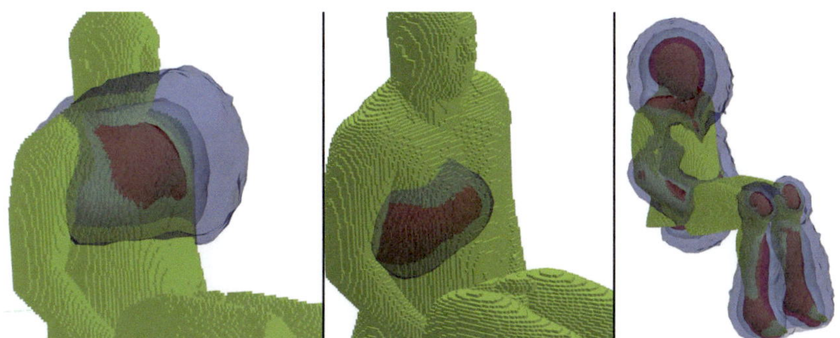

Figure 4.4: Isoflux surfaces for ^{241}Am in lungs *(left)*, liver *(centre)*, and skeleton *(right)* (Marzocchi 2011). The flux generally decreases with increasing distance from the source and increasing thickness of the surrounding tissue. Regions with high flux *(red)* likely result in high counting efficiencies for detectors positioned accordingly. The voxel phantom was cut to change its body posture.

4.3.4 Thyroid

A large portion of iodine in the human body is stored and processed by the thyroid. According to Lynch (2011), ^{125}I and ^{131}I can be detected by a detector at the neck approximately 10 cm superior the cricoid cartilage.

4.3.5 Other

Marzocchi et al. (2011) select optimal detector positions for a partial-body setup with a computational phantom in chair and stretcher position by calculating maximum photon flux around the source structure (figure 4.4), but do not specify any method or guidelines to reproduce these positions in general.

Hegenbart and Breustedt (2011) use a position recording system for a rack of two phoswich detectors with sensors related to the mechanics of the rack to track position and orientation.

5 Computational body counting

Classical body counting with physical phantoms has restrictions in measurement time, available radionuclides and spatial activity distribution, and most importantly in the body shapes and anatomy of phantoms. Moreover, it is inefficient to perform sensitivity analyses requiring several hundred measurements. Radiation transport simulation with accurate physics models of particle interactions and anatomically realistic computational phantoms enables radiation protection research to efficiently perform sensitivity analyses for computational body counting with arbitrary activity distributions.

5.1 Monte Carlo method

Radiation transport is a complex computational problem which describes propagation of ionizing radiation through matter and the accompanying energy transfer. It requires modelling of a radiation source, the general interaction processes of the particles being transported, the media in which the transport is performed, and how and where the individual particles are being tallied. Applications for radiation transport simulation are, for example, treatment planning for radiotherapy, computed tomography simulation, and reconstruction of radiological accidents (Kling et al. 2001).

One common method to solve radiation transport is the *Monte Carlo method* (Metropolis and Ulam 1949). This is a class of computational algorithms that solve problems depending on random variables with defined distributions by sampling from these distributions and averaging the resulting quantity. According to the central limit theorem (Koroliuk 2013), the mean of a sufficiently large number of independent random variables will approximately be a normal distribution. This means, that sampling the interaction events and normalizing the computed quantity to the number of source emissions can be characterized by mean and variance, which reduces with the number of samples used for the quantity.

5.1.1 Monte Carlo codes

Several general purpose Monte Carlo tools or *codes* for the application
to radiation transport exist. Some of the more established codes are
MCNP5/MCNPX (Pelowitz 2007), GEANT4 (Agostinelli et al. 2003),
EGSNRC (Kawrakow et al. 2011), PENELOPE (Salvat, Fernández-Varea
and Sempau 2011), and FLUKA (Fassò et al. 2003). From a user perspective, the codes differ mainly in the supported data formats, the physics
models, and the input and output syntax. Modelling a scenario in all
these codes requires a set of data consisting of:

Sources with probability distributions defining particle type, energy,
position and orientation of an emission.

Media with geometric models, densities, chemical compositions and
electrical conductivity for radiation transport.

Transport parameters with cross sections depending on chemical element,
particle type and energy for particle interactions.

Tallies associated with a surface or volume and a radiation quantity (e.g.
fluence, energy deposit, or pulse height).

Monte Carlo codes perform a stepwise random walk for particles
generated by the source with heavy use of a pseudo-random number
generator for sampling the associated probability distributions:

1. Check the problem termination criterion (e.g. number of computed
   particle histories, elapsed computer time, or achieved tally precision).
   If it is reached, report the tally information and stop the run.

2. If the particle stack is empty, produce a new primary particle
   according to the source definition and push it on the stack. Select
   the topmost particle on the stack.

3. Check the particle termination criterion (e.g. lower energy threshold).
   If it is reached, the particle deposits its remaining energy at its local
   position and terminates.

4. Perform a transport step that moves the particle to a new location
   considering the current medium and intermedia borders.

5. Check if the tally region has been traversed, and update the tally
   quantity if this is the case.

6. Perform a particle interaction with the current medium. Update
   the particle parameters (e.g. energy, and orientation), and produce
   secondary particles and put them on the stack. Return to step 1.

5.1.2 Variance reduction

Because of the independent and memoryless transport of particles, the
Monte Carlo method is highly suited for parallelization. Additionally,
the computational efficiency of the basic transport algorithm can be
increased by introducing variance reduction methods. These methods
reduce the variance of the computed values given the same computer time
by introducing a comparably small bias in the mean value.

Automated methods use general assumptions about particle transport
and are part of the basic implementation of the code requiring no further
modelling. One example of such a method is the combination of multiple
scattering events for electrons (Kawrakow et al. 2011).

Manual methods require information about the particular scenario, and
must be adjusted by the user. One idea is to use background information
about the modelled scenario to predict the relative expected contribution
of particles with certain characteristics to the tally value. Particles with
high contribution should be assigned a high importance regarding further
propagation and others should be terminated prematurely. This method
is called *importance sampling* (Pelowitz 2007) and is a standard method
for Monte Carlo codes.

Other variance reduction methods are directly applied when modelling
a scenario. The idea is to simplify parts of the geometric model to speed
up the simulation. A simulation may even be split into multiple parts
or stages by defining a geometric interface that spatially separates the
parts. Each stage computes the fluence of particles at its interface, which
is input to the next stage as a surface source (Pelowitz 2007).

5.2 Monte Carlo N-Particle eXtended

Monte Carlo N-Particle eXtended (MCNPX) is a Monte Carlo radiation
transport code developed at Los Alamos National Laboratory (LANL). It
is capable of simulating a large range of particle interactions and energies.
MCNPX specifies an input language that is interpreted upon execution.

After the simulation problem has been computed, the results are output in a special file format called MCTAL. The MCNPX input is divided into a list of *cards* that specify the structure and behaviour of the radiation transport. Each card has a unique (alpha-)numerical identifier and may be referenced by other cards. A full specification of the input and output formats is given by Pelowitz (2007).

5.2.1 Sources

Cards associated with the specification of sources are mainly distributions which map certain values of a quantity (e.g. particle type, energy, position, and orientation) to a probability. Several predefined functions for continuous distributions are available. These distributions can be nested to define complex sources. Emission locations can be points, surfaces, or volumes. Locations for any space are produced using an enclosing volume and rejection sampling.

5.2.2 Geometry

Geometry in MCNPX implements the *constructive solid geometry* approach (Mortenson 1985) with quadric surfaces (Lennerz and Schömer 2002). These are surfaces defined by quadratic polynomials $f(x)$ with symmetric matrix $A \in \mathbb{R}^{3 \times 3}$, vector $a \in \mathbb{R}^3$, and literal $a_0 \in \mathbb{R}$.

$$f(x) = x^\top A\, x + 2\, a^\top x + a_0 \tag{5.1}$$

A surface is defined by all points x where $f(x)$ vanishes.

$$\left\{ x \in \mathbb{R}^3 \mid f(x) = 0 \right\} \tag{5.2}$$

Volumes or *cells* are defined as logical combinations of the spaces $f_i(x) \leq 0$ separated by surfaces i. The basic logical operators are negation $\neg$, conjunction $\wedge$, and disjunction $\vee$.

$$\left\{ x \in \mathbb{R}^3 \mid \bigvee \bigwedge (\neg)\, f_i(x) \leq 0 \right\} \tag{5.3}$$

In addition, repeated structures can be defined. These are three-dimensional lattices of equivalent primitive cells that are either hexahedra or hexagonal prisms. Cell cards have associated media and specify their own densities.

5.2.3 Media

The properties of complex media relevant for radiation transport are density, chemical composition, and conductivity. The latter two are specified using material cards. Chemical composition is given using mass or atomic fractions of the individual elements or their isotopes identified by atomic number and mass number. Associated libraries specify cross sections for these.

5.2.4 Tallies

Predefined tally types are current, fluence or flux, track length, collision heating, energy or charge deposition, and pulse height in a geometric object. A binning discretizes a domain, such as energy, angle, space, or time that the quantity is related to. The reported values are given for each bin with relative error and normalized to the number of computed particle histories. After problem completion, statistical tests are performed to check for convergence of the values.

Standard output quantities (Shultis and Faw 2006) are the number of computed particle histories N, elapsed computer time t, mean value μ, and variance σ^2 for each bin. The relative error is given as $R = \sigma/\mu$.

$$\mu = \frac{1}{N} \sum_{i=1}^{N} x_i \tag{5.4}$$

$$\sigma^2 = \frac{1}{N(N-1)} \sum_{i=1}^{N} (x_i - \mu)^2 \tag{5.5}$$

Several statistical quantities are evaluated that provide additional information about the convergence of the problem. These are the figure of merit (FOM), and the variance of the variance (VOV).

$$FOM = \frac{1}{R^2 t} \tag{5.6}$$

$$VOV = \frac{1}{\sigma^2} \sum_{i=1}^{N} (x_i - \mu)^4 - \frac{1}{N} \qquad (5.7)$$

5.2.5 Variance reduction

MCNPX supports a set of automated and manual variance reduction methods using population control, modified sampling, and partially-deterministic calculations. An overview of those methods is given by Shultis and Faw (2006). And a comparison with application to body counting is given by Farah, Broggio and Franck (2011a).

5.3 Application to body counter calibration

Radiation transport simulation with the Monte Carlo method is an efficient and intuitive tool for computational body counter calibration. It gives comparable results to measurements (Broggio et al. 2012; Gómez-Ros et al. 2008; Hegenbart et al. 2009; Liye et al. 2007). Body counting scenarios have two significant components: computational phantoms with associated sources, and detectors with associated tallies. If necessary, models of the measurement chamber and scattering objects may also be created.

5.3.1 Computational phantoms

Computational phantoms are modelled in MCNPX either using quadric surfaces or repeated structures. Mathematical phantoms based on quadric surfaces can be directly converted to cell and surface cards. Voxel phantoms based on tomographic imaging are supported by repeated structures. All other geometric representations must either be approximated by quadric surfaces (which is often not viable) or *voxelized.*

Computational phantoms are sectioned into several regions with comparable tissue, functionality, and importance to the radiation transport. Any of those regions may be declared as a source. Similar to classical body counter calibration, the spatial source distribution is defined as homogeneous, and the energies are discrete values representing gamma or X-rays of radionuclides.

5.3.2 Detectors

Detectors are approximated by combinations of quadric surfaces. Many researchers use radiography and point sources in combination with Monte Carlo simulations and parametric detector models to individually characterize their detectors (Britton et al. 2012; Elanique et al. 2012; Marzocchi, Breustedt and Urban 2010; McNamara et al. 2012; Nogueira et al. 2010). Particular attention to detail is paid to entrance window and dead layer of the detector crystal as slight changes in these structures have large impact on the detection of low-energy photons.

The response of the detector crystal to an incident photon is quantified by a pulse-height tally assigned to the active volume. Since the energy calibration model (equation 3.2) is usually linear, it is simply modelled by specifying the energy range and the number of channels. Detector energy resolution is simulated with a method called *Gaussian energy broadening*. The method samples each pulse in the active detector volume with an unbiased normal distribution before recording it in the pulse-height tally. The standard deviation of the distribution is defined according to the energy resolution model (equation 3.1).

5.3.3 Modelling tools

International comparisons and training actions (Broggio et al. 2012; Gómez-Ros et al. 2008; Lopez et al. 2011) among research institutes in the field of radiation protection including modelling, simulation and evaluation tasks are common to assess the state of the art in radiation transport simulation and to implement quality assurance. Several software tools have been developed in the recent years assisting in body counter calibration, the simulation or recreation of radiological accidents, and the creation and manipulation of phantoms (Pölz et al. 2013):

VISUAL MONTE CARLO (Hunt et al. 2003) is a Monte Carlo code created for calibration of body counters, and dose calculation for internal and external radiation sources.

EGSNRCMP (Kawrakow, Mainegra-Hing and Rogers 2006) is a multiplatform environment for running the Monte Carlo code EGSNRC. It includes a graphical user interface that enables users to modify simulation settings, and also to view and edit media properties.

MCNPX VISUAL EDITOR (Schwarz 2008), SABRINA (Riper 2003b), and MORITZ (Riper 2003a) provide graphical user interfaces with three-dimensional geometry visualization for the Monte Carlo code MCNPX.

OEDIPE (Chiavassa et al. 2005) is a tool to handle voxel phantoms for body counter calibration and targeted radiotherapy. It allows for fast creation of voxel phantoms from medical imaging data, automatic association with MCNPX and processing of simulation results.

SESAME (Huet et al. 2009) is a tool to perform numerical reconstruction of radiological accidents involving external sources for simulation with MCNPX. It can create voxel phantoms with adjusted posture and morphology based on medical imaging data (Courageot, Sayah and Huet 2010).

VISUAL WORKSHOP (Bird and Fry 2013) is a visualization and analysis tool for the Monte Carlo code MCBEND (Cowan, Dobson and Martin 2013) amongst others. It can display geometric models defined in input files, and visualize tally scores from output files. It also organizes and manages all files, and provides a basic editor.

SIMPLEGEO (Theis et al. 2006) is a modelling tool specifically created to unify the various geometry modelling processes and syntaxes of Monte Carlo codes for radiation transport scenarios.

VOXEL2MCNP (Hegenbart et al. 2012) is a tool supporting users in modelling radiation transport scenarios using voxel phantoms and other geometric models, generating corresponding input for MCNPX, and evaluating simulation output. Its primary applications are body counter calibration and calculation of specific absorbed dose fractions for internal and external dosimetry.

All modelling tools require some form of data model to represent the data they are working with — both in volatile memory for data processing, and in physical memory for data storage and exchange. The available literature does not contain any notion of how those data models are designed. Presumably, they contain an implementation of the data models that the associated Monte Carlo codes are using and provide additional support for several general purpose modelling tools.

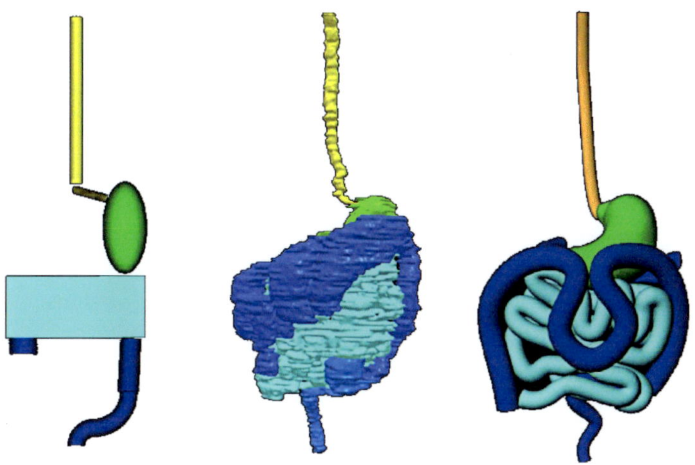

Figure 5.1: Stylized, voxel and hybrid (NURBS) models of the alimentary tract (Lee et al. 2007). The stylized model originates from the Oak Ridge National Laboratory (ORNL).

5.4 Development of computational phantoms

The state of the art in computational phantom development changed significantly over the past (Bolch et al. 2010; Segars and Tsui 2009). Development began in the 1960s with mathematical or *stylized models* based on quadric surfaces native to radiation transport codes, moved on to *voxel models* in the 1980s with the advent of medical tomographic imaging, and took advantage of the progress in computer graphics regarding surface representation methods and combined the advantages of stylized and voxel modelling to *hybrid models* with polygon meshes and non-uniform rational basis splines (NURBS) (Piegl and Tiller 1997) in the 2000s. The change in model representation methods (figure 5.1) is driven by the desire to modify and individualize phantoms while providing and maintaining anatomical accuracy. Detailed overviews of existing computational phantoms are provided by Zaidi and Xu (2007), Zaidi and Tsui (2009), and Xu and Eckerman (2009).

5.4.1 Voxel models

Voxel models are based on segmentation of medical tomographic imaging data. The method of choice is computed tomography because of its high spatial and temporal resolution. An alternative is magnetic resonance imaging, which offers higher contrast for soft tissue, but introduces motion blur. Voxel modelling defines the gold standard in anatomical realism. However, image segmentation is still a difficult problem in computer science, and the creation of large voxel phantom libraries does not seem feasible without enormous progress in segmentation algorithms.

Another major drawback of voxel models is their inherent structure, which is a fixed three-dimensional lattice of voxel elements. This only allows the application of low-order manipulation methods as applied by ICRP (2009). Voxel operations are the same as those available in basic image segmentation software, for example, dilation, erosion, region growing, thresholding, filters, and scaling (Dougherty 1992).

5.4.2 Stylized models

The stylized approach models anatomic structures with mathematically-defined surfaces. This is usually interesting for applications where only insufficient imaging data is available and anatomic background knowledge is applied (Farfán et al. 2004), or anatomical details are unimportant and being deliberately removed.

Voxel models are superior to stylized models with regard to anatomical accuracy. However, stylized models offer more possibilities for modification, because the models are usually less complex and have fewer degrees of freedom. These modifications are not more realistic than those for voxel models, because they are not based on any imaging data and still have too many degrees of freedom, but are generally easier to perform with three-dimensional geometry modelling tools.

5.4.3 Hybrid models

Hybrid models are a combination of voxel and stylized modelling. They approximate the segmented structures of a voxel model with surfaces while constraining information loss to a certain degree (Lee et al. 2010). Hybrid phantoms preserve both the anatomical realism of voxel models and the structural flexibility of stylized models. However, anatomically realistic modifications of computational phantoms can only be performed by using statistical imaging data and appropriate geometric representation formats.

- Lee et al. (2007) use graphic modelling tools to approximate voxel models with NURBS meshes. Segmentation errors are corrected using existing organ models and anatomical background knowledge. Additionally, the posture can be changed to some extent.

- Mofrad et al. (2010) use spherical harmonic functions to approximate the surfaces of 35 livers from computed tomography data to create a statistical shape model. The model is instantiated into anatomically realistic shapes by specifying the desired distance from the mean shape in the principal components.

- Segars and Sturgeon (2010) create computational phantoms from coarsely segmented computed tomography data by performing volumetric registration with a hybrid phantom.

- Kim et al. (2011) use graphic modelling tools to approximate a voxel model with a mixture of polygon and NURBS meshes.

Examples of the commonly used modelling tools are IMAGEJ (Ferreira and Rasband 2012) for image series and voxel lattices, 3D-DOCTOR (Able Software Corp 2013) for segmentation and surface approximation of DICOM data, and RHINOCEROS (Robert McNeel & Associates 2012) and BLENDER (Stichting Blender Foundation 2013) for modelling with curved surfaces.

6 Personalisation methods

The methods available in literature for the personalisation of calibration factors for body counter calibration can be categorized into *adaptation-based* or *interpolation-based* methods. Adaptation-based methods (Henriet et al. 2012) select a phantom similar to the individual from a case base, and modify the selected phantom to increase its similarity. Interpolation-based methods (Doerfel, Heide and Sohlin 2006; Henriet et al. 2012; Lynch 2011; Mohr and Breustedt 2007; Pierrat et al. 2007) perform calibration on a series of phantoms, and create an estimator based on samples of counting efficiency. Both approaches rely on sensitivity analyses of sample calibration data to identify structures of the human anatomy relevant for the particular measurement setup. To evaluate these different approaches, a closer look at methods to create phantom series with varying anatomical features is necessary.

6.1 The reference man paradigm

Many researchers have created computational phantoms over the past decades (Xu and Eckerman 2009; Zaidi and Tsui 2009; Zaidi and Xu 2007) for application in internal dosimetry, medical imaging simulation, radiotherapy and interventional radiology, and applications involving non-ionizing radiation. The main differences in computational phantoms are due to different requirements of the simulation (e.g. radiation, thermodynamic, or biochemical transport properties), different scales (e.g. macro, micro, nano), and the desired degree of individualization versus the available data (imaging data or anthropometric parameters). The degree of phantom individualization in body counting is restricted by the available data, which is primarily from external body measurements or information acquired by interviewing the person. Therefore, computational phantoms for body counting are mostly phantoms representing the average of a population of individuals.

Accordingly, ICRP (2002) specifies the reference men and women of different age groups with basic anatomical and physiological data for use in radiation protection. The anatomical data includes anthropometric parameters (e.g. body mass and height), organ masses and morphology (e.g. bone surface area and length of alimentary tract segments), and organ and tissue media properties (e.g. density and chemical composition). The reference man specification is part of a paradigm (ICRP 2007) in radiation protection, which calculates the dose to the reference person in the individual exposure scenario instead of to the individual. It is assumed that if the reference person is protected (i.e. doses are below defined limits), also the majority of individuals is protected.

Many researchers (Cassola et al. 2010; ICRP 2009; Lee et al. 2010; Zhang et al. 2009) have implemented the 2002 reference man specification by creating computational phantoms with matching body height, body mass, and organ masses. The main application for those phantoms is the calculation of organ absorbed dose fractions for internal and external radiation sources.

6.2 Series of computational phantoms

Having developed computational phantoms, many researchers make modifications to create entire phantom series using anthropometric parameters like body mass, body height, chest circumference, and breast size (an example is given in figure 6.1). These phantom series are used to perform sensitivity analyses for various applications.

6.2.1 Morphometric categories

Phantoms can be classified (Bolch et al. 2010) depending on the number of people they are intended to represent or rather the amount of data of different people that were used to create them:

Person-specific phantoms are directly based on one person and have the inherent anatomical characteristics of that person. A common modelling method is segmentation of tomographic imaging data to create voxel phantoms.

Reference phantoms are the average of a large group of people. A form of stylized modelling is usually involved in creating reference phantoms.

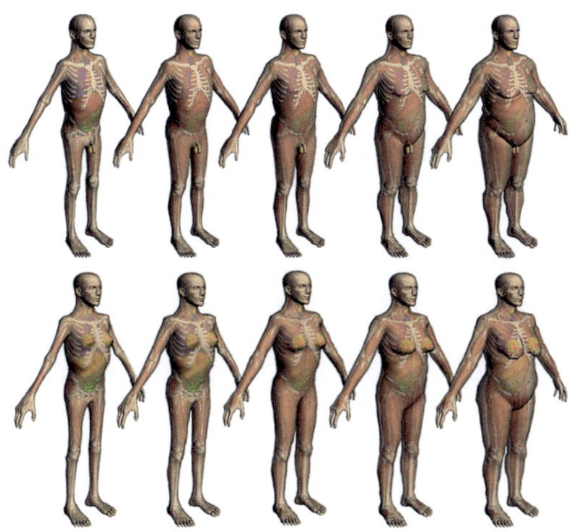

Figure 6.1: Series of adult male and female computational phantoms with reference body height and varying body masses representing 5th, 25th, 50th, 75th and 95th percentiles of variation in the U.S. population (Na et al. 2010).

Person-dependent phantoms are between both extremes. They are used when no person-specific imaging data is available and reference phantoms are not appropriate for the application. They are modelled either by generalizing person-specific phantoms or by specializing reference phantoms by removing or adding anatomical details. Most hybrid phantoms are person-dependent.

A common way to perform specialization is to use statistical data of body mass and height from health and nutrition surveys to derive population percentiles in a particular range of these quantities. Starting from a reference phantom, person-dependent phantoms are generated representing averages of population groups:

- Johnson et al. (2009) create a series of person-dependent phantoms for application to dosimetry with varying body height, body mass, and various circumferences by scaling, NURBS modelling, and changing adipose tissue volume from age-dependent reference phantoms of the UFH series.

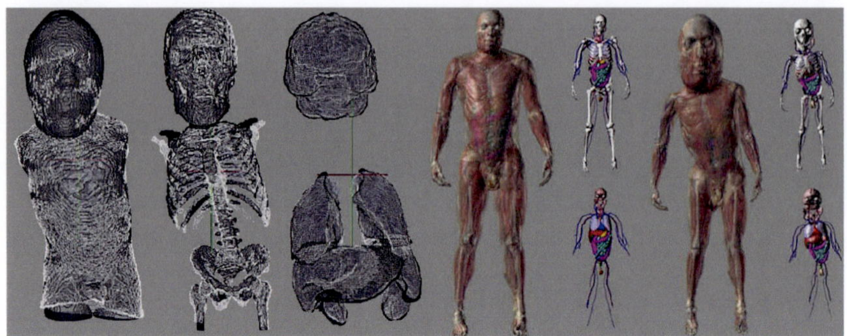

Figure 6.2: Computational phantom modification by volumetric registration of a coarsely segmented data set *(left)* with a reference phantom *(centre)* to create a person-specific phantom *(right)*. The segmented structures are body surface, skeleton, brain, lungs, liver, spleen, stomach and kidneys (Tward et al. 2011).

- Na et al. (2010) create a series of person-dependent adult phantoms for application to external dosimetry based on RPI-AF and RPI-AM with varying body height, body mass and organ masses. They increase or decrease the volume of structures and resolve overlap by deforming other nearby structures.

- Cassola et al. (2011) create a series of person-dependent standing adult phantoms for application to external dosimetry based on FASH and MASH with varying body height, body mass and organ masses. The modifications are done with NURBS modelling tools.

- Segars and Sturgeon (2010) take a completely different approach by creating a series of person-specific phantoms from coarsely segmented computed tomography data by performing multi-channel large deformation diffeomorphic metric mapping with a high-detail hybrid reference phantom for application to medical imaging simulation (figure 6.2. Missing details in the segmentation are mapped from the reference phantom by using the relative change in positions of the associated structures between both phantoms.

6.3 Interpolation-based methods

Anthropometric parameters are quantities that quantify anatomical features. *External parameters* are acquired with simple measurement equipment on the body surface, such as body circumferences, lengths, and skinfold thicknesses. *Internal parameters*, on the other hand, require insight in the internal structures of the body and cannot be acquired easily. Examples are organ volumes, tissue thicknesses and shape classifications. Obviously, internal parameters offer more information about the structures of the human body that are relevant for radiation transport, but their measurement usually requires some kind of medical imaging.

The current state of the art for personalizing calibration factors (Doerfel, Heide and Sohlin 2006; Lynch 2011; Mohr and Breustedt 2007; Pierrat et al. 2007) in lung and liver counting is based on an internal anthropometric parameter called *chest wall thickness*. Personalisation methods for other counting scenarios are not available in literature.

6.3.1 Chest wall thickness

The transmission Φ/Φ_0 of a narrow photon beam through the chest wall (the tissue between lungs and chest surface or liver and surface of the upper abdomen) can be expressed by according tissue thickness x_{cw} and linear attenuation coefficient μ_{cw}. For a fixed photon energy, the counting efficiency η of a detector covering a certain portion of the chest wall is proportional to the average photon transmission through this tissue. For increasing photon energy, the portion of photons interacting with the detector decreases (equation 4.3), which is not included in the following relation.

$$\eta \propto \frac{\Phi}{\Phi_0} = \mathrm{e}^{-\mu_{\mathrm{cw}}\, x_{\mathrm{cw}}} \tag{6.1}$$

Since the chest wall contains layers of muscle tissue, adipose tissue, bone, and cartilage, μ_{cw} is a complex property. Due to the higher absorption of bone and cartilage, the chest wall is nearly opaque to low-energy photons at the ribs. Therefore, chest wall thickness is usually only expressed for ratios of muscle tissue w_{m} and adipose tissue $w_{\mathrm{a}} = 1 - w_{\mathrm{m}}$ at the intercostal spaces.

$$\mu_{\mathrm{cw}} = w_{\mathrm{a}}\,\mu_{\mathrm{a}} + w_{\mathrm{m}}\,\mu_{\mathrm{m}} \tag{6.2}$$

6.3.2 Measurement of chest wall thickness

Chest wall thickness is measured using medical imaging equipment. If tomographic imaging data is available, the thickness and composition is assessed using image segmentation. An alternative is ultrasound (Gün 2010; Lynch 2011; Sumerling and Quant 1982).

The transmission is measured at several points at the intercostal spaces covered by the detector (depending on the size of the entrance window) and averaged. To have an intuitive and comparable value, the average is then typically normalized to reference muscle tissue μ_m. The resulting quantity is called muscle-equivalent or *effective chest wall thickness* (Sumerling and Quant 1982).

$$\hat{x}_\mathrm{cw} = -\frac{1}{\mu_\mathrm{m}} \ln \frac{1}{n} \sum_{i=1}^{n} \mathrm{e}^{-\mu_i x_i} \tag{6.3}$$

Several authors have also measured chest wall thickness for computational phantoms (Hegenbart, Gün and Zankl 2010; Kramer, Hauck and Allen 2001).

An alternative to the complex and time-consuming measurement is the estimation of chest wall thickness using external parameters. For example, chest wall thickness is correlated to the ratio of body mass and body height, and body mass index. Population-specific formulas are used for application to body counters.

6.3.3 Personalisation method

The current methods for creating personalised calibration factors in lung and liver counting combine a calibration model and a person model. The person model f describes chest wall thickness of a person dependent on body mass m and body height h. This is usually a linear model of m/h or m/h^2 based on tomographic imaging data or ultrasound measurements. Since this data is typically derived from published proband studies, adjustment of the chest wall thickness to a specific measurement setup is not possible.

$$\hat{x}_\mathrm{cw} = f(m, h) \tag{6.4}$$

The calibration model g describes counting efficiency dependent on chest wall thickness. This model is usually a linear interpolation of samples acquired from a physical torso phantom with a lung or liver set with source s emitting photons of energy E, detector t, and different chest overlays to modify thickness and muscle-adipose tissue ratio of the chest wall.

$$\hat{\eta}_{t \leftarrow s}(E) = g_{s,t,E}(x_{\mathrm{cw}}) \tag{6.5}$$

6.3.4 Cup size

Lung counting for females is more difficult, since part of the breasts add additional attenuation, and the variability of cup size is relatively high. This topic was addressed by several authors. Hegenbart et al. (2008) and Farah, Broggio and Franck (2010) created series of person-dependent female phantoms with varying cup sizes for application to lung counting with phoswich and germanium detectors respectively. They found a negative correlation of breast mass and cup size to counting efficiency.

6.4 Adaptation-based methods

Adaptation-based methods directly modify computational phantoms to match the individual. Only one adaptation-based method for application to body counting was found in literature. Henriet et al. (2012) present the EQUIVOX framework based on *case-based reasoning* for lung counting. Case-based reasoning is a class of algorithms that uses a set of problems with known solutions to construct solutions for new problems. The idea is that similar cases have similar solutions and only small changes are needed if a large case base covering the expected domain of problems is available. The basic case-based reasoning algorithm has four main processes: retrieve, reuse, revise, and retain.

The case base is a set of problems with an associated solution generated by experts. EQUIVOX uses 24 adult female computational phantoms with varying cup size and chest girth based on ICRP-AF and NURBS modelling (Farah, Broggio and Franck 2010).

Retrieve: Given a new case, retrieve all cases from the case base similar to that case. This requires a similarity metric. EQUIVOX describes each case with a feature vector, consisting of, for example, age, body mass, body height, gender, chest girth, and underbust girth. It computes a similarity score relating the new case to each existing case. A set of cases satisfying a constraint (e.g. minimal or below a certain threshold) on this score is selected.

Reuse: Given a set of cases and their solutions, create a solution for the new case. EQUIVOX describes each known case for the purpose of adaptation with polygonal surface mesh of the lungs consisting of a fixed number of vertices. *Artificial neuronal networks* (ANN) are used to interpolate between two meshes dependent on body height.

Revise: Check if the constructed solution actually solves the new case and make changes if necessary. This usually requires some kind of user interaction. EQUIVOX revises a case with help of experts by checking if the generated mesh matches the lungs according to tomographic imaging data of the person.

Retain: Add the new case and its solution to the case base. EQUIVOX adds the solved case to the database if tomographic imaging data was available and the case was properly revised.

7 Statistical analysis

Statistical learning or *machine learning* (Hastie, Tibshirani and Friedman 2009; MacKay 2003) is a theory for the construction of systems that learn from data. It plays an important role in computer science (e.g. data mining, computer vision, natural language processing, and information retrieval), and related fields.

Supervised learning is a branch of machine learning that uses discrete observations to estimate an unknown, target function that relates a set of features to a target variable. The discrete version of supervised learning is called *classification*, and the continuous version is called *regression*. Regression is either *parametric* or *non-parametric*. This specifies whether the underlying model of the target function is explicitly or implicitly defined.

7.1 Risk minimization

Given a target function $f\colon X \to Y$ with $f(x) = y$ and a hypothesis $h\colon X \to Y$ that estimates f with $h(x) = \hat{y}$. The distance between h and f at x is described by a loss function $r\colon Y \times Y \to \mathbb{R}$ with $r(y, \hat{y})$. The *risk* $R(h)$ of the hypothesis is the sum of all losses with regard to the probability density $P(x)$ of X.

$$R(h) = \int_{x \in X} r(y, \hat{y}) \, \mathrm{d}P(x) \tag{7.1}$$

The goal of regression is to find $\hat{h}$ in a set of hypotheses $\mathcal{H}$ that minimizes the risk.

$$\hat{h} = \arg \min_{h \in \mathcal{H}} R(h) \tag{7.2}$$

In applications, the probability distribution of X is unknown, but it is possible to observe $f(x)$ on locations sampled from $P(x)$. Let

$\{(x_i, y_i) \mid i = 1, \ldots, n\}$ be a set of discrete observations of f with uncorrelated, unbiased errors ε_i. Moreover, it is assumed that the x_i can be measured without error. Otherwise an errors-in-variable model (Carroll et al. 2006) is required.

$$f(x_i) = y_i + \varepsilon_i \tag{7.3}$$

Without further knowledge about $P(x)$, a simple approximation of the risk based on the observations is the *empirical risk* $R_\mathrm{E}(h)$. It describes how well the estimates $\hat{y}_i$ represent the observations y_i at x_i.

$$R_\mathrm{E}(h) = \frac{1}{n} \sum_{i=1}^{n} r(y_i, \hat{y}_i) \tag{7.4}$$

Any hypothesis that interpolates the data minimizes empirical risk. To account for that, the hypotheses space must be restricted. The classical approach for function regression uses a parameterized function series and gradient descent for optimization. If no structural assumptions about the target function can be made, a non-parametric approach must be taken. The two main non-parametric methods are *local regression* and *structural risk minimization*.

7.2 Local regression

Local regression methods (Cleveland and Devlin 1988) are based on the idea that the closer points are to each other in the feature space, the more likely they are related with a simple functional. That functional is usually a parametric polynomial $h_\beta(x)$ of low degree d. It may even be a constant.

$$h_\beta(x) = \sum_{j=0}^{d} \beta_j\, x^j \tag{7.5}$$

Instead of performing a global parametric fit, each point is fit locally. The result is a function with dynamic parameters, which can also be written as the weighted sum of target values in the neighbourhood of x. The dynamic weights w_i only depend on the feature values of the samples, but not their target values.

$$h_\beta(x) = \sum_{j=0}^{d} \beta_j(x)\, x^j = \sum_{i=1}^{n} w_i(x)\, y_i \tag{7.6}$$

The neighbourhood of x is determined by *smoothing kernels* $K(x)$. This gives more weight to points near x and less weight to those further away. The distance between two points is determined by a distance metric d (e.g. Euclidean, Chebychev, or Mahalanobis distance). λ is the bandwidth of the kernel. It can be constant or adaptive, for example, to include a fix number of neighbouring points relative to the total number of observations.

$$K(x, x_i) = K\left(\frac{d(x, x_i)}{\lambda(x)}\right) \tag{7.7}$$

Cleveland (1979) describes a set of restrictions for smoothing kernels, which allows for a large range of functions. Common types of kernels are Triweight, Epanechnikov, or Uniform functions.

The risk $R_{\mathrm{LP}}(h)$ at x is then expressed as the sum of the kernel-weighted squared residuals. Its minimization is a weighted least squares problem.

$$R_{\mathrm{LP}}(h) = \sum_{i=1}^{n} K(x, x_i)\, (y_i - \hat{y}_i)^2 \tag{7.8}$$

Cleveland (1979) proposes an iterative method for locally weighted polynomial regression leading to robust estimators. For each iteration, the kernel weights are adjusted with robustness weights dependent on the deviation of the residuals of the current hypothesis from their mean. This ensures that the residuals are unbiased and have low maximum variation.

7.3 Structural risk minimization

With increasing complexity of hypotheses, the size of the hypotheses space increases because of the additional degrees of freedom. The idea of structural risk minimization (Vapnik 1999) is that if multiple hypotheses fit the data with similar empirical risk, the hypothesis with minimal complexity should be favoured, because it is the most probable one. This idea is closely related to the concept of *minimum description length* (Rissanen 1978) in information theory. A regularization function describing hypotheses complexity $R_\mathrm{S}(h)$ is modelled with an inner product $\|\cdot\|_\Omega$. An example is the Vapnik-Chervonenkis dimension (Vapnik and Chervonenkis 1971), which provides a probabilistic upper bound on the risk.

$$R_\mathrm{S}(h) = \|h\|_\Omega^2 \tag{7.9}$$

The risk $R_\mathrm{SR}(h)$ of a hypothesis is therefore described as the weighted sum of the empirical risk and a regularization term. λ controls the tradeoff (Hastie, Tibshirani and Friedman 2009) between both (figure 7.1). In the terminology of estimators of random variables, the empirical risk is called *bias* (it describes the bias to the sample data), and the remaining part is called *variance* (it describes the generalization of the hypothesis to unknown data).

$$R_\mathrm{SR}(h) = R_\mathrm{E}(h) + \lambda R_\mathrm{S}(h) \tag{7.10}$$

The general solution $\hat{h}$ of minimizing equation 7.10 has the form of a linear combination of kernels $K(x, x_i)$ centred on each sample x_i with weights depending on the target value.

$$\hat{h}(x) = \sum_{i=1}^{n} w_i\, K(x, x_i) \tag{7.11}$$

An implementation of structural risk minimization for regression is nonlinear *support vector regression* (Drucker et al. 1997). This method uses an implicit mapping $\varphi \colon X \to \mathcal{H}$ from the feature space to a Hilbert space with high dimension of linear functions, in which the problem is reduced to a convex optimization problem.

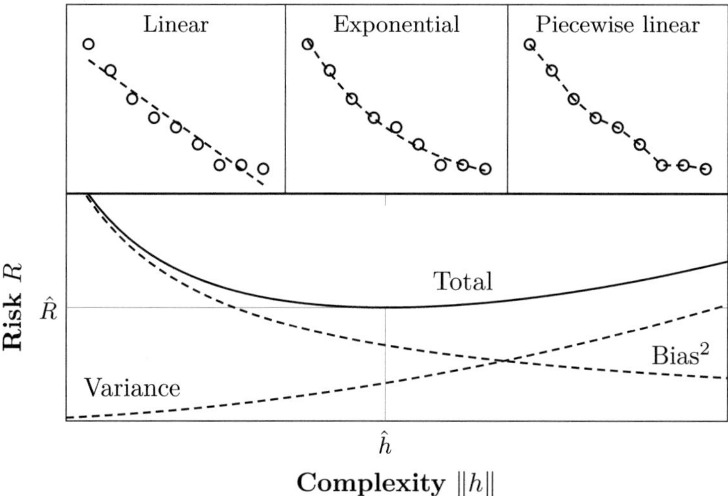

Figure 7.1: Bias-variance tradeoff and hypothesis complexity demonstrated on sample data and hypotheses. The squared bias decreases with increasing hypothesis complexity, but that also increases the variance. Too low complexity leads to *underfitting (left)*, and too high complexity leads to *overfitting (right)*. An optimal complexity $\hat{h}$ *(centre)* that minimizes the total risk (the sum of variance and squared bias) can be found. Based on Hastie, Tibshirani and Friedman (2009).

8 Meta optimization

Non-parametric regression methods have several free parameters that must be tuned according to the specific problem. They are related to the bias-variance tradeoff (Hastie, Tibshirani and Friedman 2009) and specify the smoothness of the hypothesis with regard to the data. *Meta optimization* is a collection of methods to find an optimal value for these parameters. Many meta optimization methods are based on combining *ensembles* of hypotheses (Arlot and Celisse 2010; Breiman 1994; Breiman 1996; Schapire 2003) from different subsets generated by *sampling* (Yu 2003) of the available observations to perform additional regularization of the hypothesis and make better use of the limited amount of data.

Another option to reduce the hypotheses space (besides structural risk minimization) is the reduction of the number of features that are provided in the sample data. This approach, called *feature subset selection* or *feature selection*, is based on the assumption that many features are redundant among each other or irrelevant to the target function. Important applications for feature selection are the analysis of DNA microarrays for prediction of various health conditions (Guyon et al. 2002; Mukherjee et al. 1999), text filtering (Bekkerman et al. 2003; Dhillon, Mallela and Kumar 2003), and face recognition (Gundimada, Asari and Gudur 2010; Yang et al. 2007).

8.1 Performance measures

Similar to empirical risk, there are also measures that are more suitable for meta optimization. These measures are primarily based on residuals $r_i = |y_i - \hat{y}_i|$ or on changes in ranks $r_i = p_i - \hat{p}_i$ with $p_i = |\{y_i \leq y_j \mid j = 1, \ldots, n\}|$ between target function and hypothesis, given observations $y = (y_i)_{i=1,\ldots,n}$ of the target value, and their estimates $\hat{y} = (\hat{y}_i)_{i=1,\ldots,n}$.

Common residual-based measures are *root mean squared error, root relative squared error,* and *Pearson's product-moment correlation.*

Ranking-based measures (Kendall and Gibbons 1990) describe how well the hypothesis reflects the true order of the values. They are robust to outliers. Common examples are *Kendall's* τ and *Spearman's* ρ.

8.2 Resampling methods

A simple way to reduce bias is to split the set of observations into *training data* and *test data*. Regression is performed on the training data, but the performance measure is applied only to the test data. This introduces "unknown" data into the process.

Sampling methods are required to split data sets. It is important that the generated sets are representative of the distribution of the original data set. The following enumeration contains a list of popular sampling methods. A full overview of those methods and a discussion on advantages and disadvantages of resampling is given by Yu (2003).

Stratified sampling analyses the distribution of values in the data and samples accordingly to ensure that each value is represented proportionally to its observed frequency.

Bootstrap sampling performs sampling with replacement. This approximates the underlying distribution better than sampling without replacement for low sample sizes.

Cross sampling splits the data into k folds of equal size. One fold is selected as test data set, while the remaining $k-1$ folds are combined to a training data set. This process is repeated k times and each fold is selected once as test data set. This way, each data point is guaranteed to be selected for training and testing. Leave-one-out cross-sampling is cross sampling with fold size of one. This is commonly used to compute sample means and variances.

Random sampling repeatedly and randomly splits the data into subsets of defined size.

8.3 Ensemble learning

The idea of ensemble learning is that an ensemble of hypotheses produces better and more robust estimates. Each hypothesis may be generated by different regression methods, by the same method on resampled data sets, or both. Ensemble learning methods define how those hypotheses are combined.

Bootstrap aggregating or *bagging* (Breiman 1994) averages hypotheses trained from bootstrap samples of the main data set with the same regression method.

Cross validation (Arlot and Celisse 2010) averages hypotheses trained from cross-sampled subsets of the main data set with the same regression method. This is often used to determine the performance of a hypothesis in combination with a simple error measure.

Stacked regression or *stacking* (Breiman 1996) combines hypotheses from different regression methods.

Additive regression or *boosting* (Schapire 2003) iteratively trains a new hypothesis on the residuals of the previous hypothesis.

8.4 Search strategies

Since many performance measures are not convex and may contain local optima, finding an optimal hypothesis is computationally hard and it is usually not viable to perform an exhaustive grid search of a large part of the space. This means that assumptions about the structure of the hypotheses space with regard to the performance measure are necessary that imply a heuristic search strategy.

Search strategies select a new hypothesis based on the sequence of past hypotheses and their performances starting from an initial guess of a good hypothesis. Different search strategies keep a varying population of hypotheses or follow a trajectory while keeping a certain number of past hypotheses, some are guided by a performance measure or have a learning component, and some greedily follow the best local decisions or perform backtracking.

Genetic algorithms (Goldberg 1989) mimic the process of natural selection by keeping a population of hypotheses that evolve by selective reproduction causing genetic crossover and mutation.

Simulated annealing (Ingber 1989) performs a random walk on the search space with the probability of moving to a worse hypothesis reducing with time. This allows leaving local optima.

Tabu search (Glover and McMillan 1986) performs a local search by checking for hypotheses that are similar to the current hypothesis and provide similar or better performances. A certain number of past hypotheses are banned to avoid cycles in local optima and ignored until their ban is lifted.

8.5 Feature selection

Feature selection (Guyon and Elisseeff 2003) is based on the concept of *minimum redundancy maximum relevance*. Redundancy refers to performance loss due to correlation of features when combining them into a subset for regression. Obviously, the combination of two features will always be less predictive regarding the performance measure than the sum of the individual performances. Relevance refers to the predictive quality of a feature subset due to correlation with the target function.

From an univariate point of view, good features are uncorrelated to other features, and highly correlated to the target function. However, this is not transferable to the multivariate case, where the combination of two features with high redundancy and low relevance may produce a perfect estimator. This is what makes feature selection so difficult.

Feature selection is a discrete optimization problem on the number and type of features that would provide the optimal hypothesis after regression analysis. There are three basic approaches: filter, wrapper, and embedded methods. Different classes of performance measures, search strategies, resampling methods, and ensemble learning are usually applied.

8.5.1 Filter methods

Filter methods embrace the minimum redundancy maximum relevance approach and try to approximate these values with ranking-based measures independent of the choice of estimator. Those measures are usually based on statistical tests and mutual information criteria.

For continuous domains and univariate cases, relevance and redundancy can be described with the F-statistic and the product-moment coefficient.

Another notion of redundancy is mutual information, which relies on empirical estimates of the probability densities of each variable. These are calculated with *kernel density estimation* which is basically regression analysis on a special domain.

Filters are computationally efficient and can be done as a preprocessing step. Their predictive power is worse than other methods, because they are independent of the actual predictor.

8.5.2 Wrapper methods

Wrapper methods (Kohavi and John 1997) use the method that is designated for the regression step to assess performance of feature subsets according to their prediction accuracy.

Compared to other methods, wrappers are computationally expensive, but likely to provide better results that are tailored to the regression method.

8.5.3 Embedded methods

Embedded methods weight each feature, which transforms feature selection into a continuous optimization problem, which can be solved with regression methods. This adds an additional layer of regression on top of the actual regression problem and allows for a high computational efficiency and prediction accuracy.

Popular embedded methods are the L_0-norm or L_1-norm support vector machine (Weston and Elisseeff 2003). The use of the according norm in the regularization term minimizes the number of features (L_0) or the sum of their weights (L_1) respectively. The support vectors (features with non-zero weights) form the optimal feature subset.

Part II

Development

9 Analysis of the state of the art

Development of an improved personalisation method for computational body counter calibration requires the analysis of modelling methods for body counting scenarios including phantoms and phantom series, detectors, software tools, and data models. The application of sophisticated methods for statistical analysis is the next step regarding sensitivity analyses leading to a new concept for interpolation-based personalisation combining the results of computational body counting and anthropometry.

9.1 Modelling

Monte Carlo methods are state of the art in body counter calibration (chapter 5). The according transition from physical to computational phantoms opens many possibilities for sensitivity analyses. The accuracy of those simulations is very much dependent on the level of detail of the associated models. So, the main focus of research is to provide methods for the reproducible creation of detailed phantom and detector models. (Other parts of research are mainly concerned with physics models, for example, regarding low-energy particle interaction effects.)

9.1.1 Detector modelling

Optimization of detector models has been successful using collimated point source measurements and radiography to estimate dead layer thickness at the electrode contact and to identify structural details in the interior of the casing (section 5.3). This is of particular importance for low-energy photons that are likely to be absorbed by those structures. Other elements of detector characterization and calibration are analogous to classical body counting (section 3.2).

9.1.2 Phantom modelling

Hybrid modelling introduces the possibility to adjust the tradeoff between the required amount of accuracy and flexibility in phantom development (section 5.4). This leads to an impressive growth in the number of computational phantoms in radiation protection research.

The problem, however, is that the requirement of an appropriate amount of imaging data is often neglected. For example, ICRP-AM (ICRP 2009) is a modification of a person-specific phantom using reference anthropometric data. It should be kept in mind when applying the phantom to a specific problem that its anatomy is not an average of the population. Other phantom developers use databases with geometric body or organ models with (presumably) non-reference shapes. They combine these individual models mainly with graphic modelling tools or morphing algorithms. This means that the anatomical accuracy of the result is solely based on the anatomical background knowledge of the modeller or the data that has been incorporated into the algorithm.

In conclusion, imaging data is paramount to create realistic phantoms, which are necessary for realistic calibration factors in body counting. Person-specific phantoms, if available, should be preferred for performing sensitivity analyses.

9.1.3 Detector positioning

Detector positioning for partial-body setups in application to routine monitoring is based on experience of technicians due to missing standards and guidelines, and high customization of body counting facilities. For computational body counter calibration and validation purposes, an accurate reconstruction of measurements and reproduction of counting setups is necessary due to high sensitivity of the relative detector locations.

This work is concerned about the comparability of simulation scenarios with each other for sensitivity analyses. For this purpose, a common (at least site-specific) positioning guideline must be developed. An intuitive method is the positioning relative to palpable, bony landmarks of the human body (section 4.3), which reduces the role of experience and anatomical expertise of the technicians.

9.1.4 Software tools

The calibration of large phantom series requires a certain amount of efficiency and reproducibility in modelling and computation. Many efforts have led to the development of software tools (section 5.3) that assist in modelling, simulation, and evaluation of radiation transport scenarios.

The software tool VOXEL2MCNP (Hegenbart et al. 2012) developed at KIT is based on the idea of input file generation for the Monte Carlo code MCNPX. It is an ideal starting point for this work, because it already provides several features for body counting for the legacy phoswich detector system at KIT. Although, extensions must be made to support other detector systems and large phantom series.

9.1.5 Data models

Exchange of phantoms, detectors and other models is common among research institutes in the field of radiation protection. Phantoms are often provided in form of binary files in custom format associated with documents containing informal text or tables. There is no standardization regarding these file formats and judging by current publications this is an inactive area of research.

Some form of standardization was introduced into the radiation protection community by the International Network of Nuclear Structure and Decay Data Evaluators (NSDD) and the Decay Data Evaluation Project (DDEP) under the authority of IAEA (section 2.1). With the distribution of the reference computational phantoms (ICRP 2009), the ICRP missed the chance to provide data in formal and semantically structured formats, giving an incentive to standardization.

The provision of structured data models for phantoms, media and annotations would be very welcome in the radiation protection community regarding the inflation of phantoms and software tools.

9.2 Personalisation

Personalisation methods for body counter calibration (chapter 6) are based on knowledge about the sensitivity of counting efficiency with respect to certain anatomical structures. This knowledge is based on sensitivity analyses using series of computational phantoms and anthropometric parameters. The use of anatomically accurate and comparable phantoms is very important for this work.

9.2.1 Development of phantom series

Many phantom series have been developed (section 6.2) and are under development for radiation transport problems and sensitivity analyses of radiation protection quantities. Manual or algorithmic modification of phantoms based on anatomical background knowledge to create variations in anatomy is a complex task which can lead to biased and unrealistic phantoms. The modification of single anthropometric parameters or anatomical structures while maintaining anatomical accuracy is very challenging. This has only been achieved for female breasts, which are due to their prominence an exception to this notion. In general, these modifications neglect the correlation of various anthropometric parameters in the population. Especially person-dependent phantoms, which have the inherent anatomy of a base phantom, are not appropriate for this application, and correlating a single parameter change to a change in counting efficiency for those types of phantoms will likely give misleading relations.

More phantoms based on imaging are created than ever before. Increasingly complex shape descriptors are applied because of their higher flexibility in phantom modification. Statistical shape models (Mofrad et al. 2010) are a very interesting concept. With an appropriate amount of sample data, they create realistic shapes ranging from population average to any extreme. But the integration of individual organs into existing phantoms remains a difficult task. Also, this approach requires extensive image segmentation to provide sufficient sample data.

The XCAT series (Segars and Sturgeon 2010) is a collection of person-specific phantoms directly based on medical imaging data. It offers high anatomical detail and accuracy, and contains a reasonable number of samples. It is therefore an ideal candidate for statistical analysis of body counting scenarios.

9.2.2 Personalisation methods

Only a few publications regarding personalisation of calibration factors exist and they are restricted to lung and liver counting. Two main approaches have been identified: adaptation-based and interpolation-based.

The adaptation-based method EQUIVOX by Henriet et al. (2012) (section 6.4) is an elaborate form of integrating the lung shape of an individual in an anatomically similar phantom by moving and adjusting the neighbouring organs and structures. The implemented case-based reasoning approach is an expert system, which improves with the number of available cases. The initial case base was constructed from the anatomy of a single phantom (ICRP-AF). Different anatomies are added with the application of the framework.

The implementation of the phantom retrieval and lung adaptation steps are kept relatively simple. The weights used in retrieval are reasonable, but not founded on their direct quantitative relation to counting efficiency. The adaptation step is solely based on body height. But, the system can compensate this in the review step in which manual adaptation of the intermediate lung shape to imaging data is performed under supervision of experts.

An option to estimate uncertainties due to different lung shapes would be to combine EQUIVOX with statistical shape models. However, the main problem here is the review step requiring extensive user interaction.

With the current state of the art in image segmentation, an interesting use of tomographic imaging data is the application of volumetric registration with a reference phantom as described by Segars and Sturgeon (2010). This method produces a similar result to the combination of the adaptation and review step of EQUIVOX, but allows to segment additional structures to improve the accuracy of the phantoms in these regions.

The presented interpolation-based method (section 6.3), on the other hand, is an interesting approach requiring little and readily available information about the individual. The entire method is focused on chest wall thickness and its correlation to counting efficiency and body weight and height. The main disadvantage of the method is the use of separate data sets, where one is based on a single calibration phantom with chest overlays and the other is not related to the body counter calibration process.

9.2.3 Statistical analysis

Machine learning provides powerful methods for regression analysis (chapter 7), for example, local polynomial regression and support vector regression. These are non-parametric methods (no model of the target function is required) with tuning parameters (e.g. kernel type or smoothing parameter). The adjustment of those parameters to a specific problem is possible using meta optimization methods (chapter 8). A very important method for meta optimization is feature subset selection. It basically performs an automated sensitivity analyses of the features with regard to the target value and provides a subset that is optimally suited for training with the desired regression method.

Filters are computationally efficient, but lack in accuracy. Wrappers are on the opposite end of the scale. They are also very simple to apply, since they consider the regression method as a black box. Embedded methods provide an elegant solution that combines the advantages of filters and wrappers.

The performance of a method largely depends on the problem, and most importantly on the quality and quantity of the available data. Performing feature selection on top of regression requires training, test, and validation sets, for which not enough data might be available.

10 Personalisation framework

The goal of this work is to quantify and reduce uncertainties in activity assessment with body counting due to variation in human anatomy to improve dose estimates for individuals exposed to radiation from radionuclide intake. The personalisation framework "STEP" was developed to achieve this goal. It is based on computational body counter calibration with phantoms representing a broad range of variation in human anatomy, and the quantification of anatomic features with anthropometric parameters. The subsequent statistical analysis creates estimators that predict counting efficiencies for specific measurement setups given a set of anthropometric parameter values. These results may be computed on demand or stored in a precomputed database. Guidelines for technicians performing body counter calibration and measurement of anthropometric parameters need to be defined to ensure reproducibility of the results for application of the framework.

10.1 Concept

The currently available interpolation-based personalisation methods (section 6.3) for lung and liver counting are based on the relation of counting efficiency η, chest wall thickness x_{cw}, and body mass m and height h for a measurement setup $t \leftarrow s$ and photon energy E. They use two linear models based on anthropometry of persons and calibration of a physical phantom (figure 10.1). The concatenation of both models forms a simple relationship:

$$\hat{\eta}_{t \leftarrow s}(E) = (g_{s,t,E} \circ f)\,(m, h) \tag{10.1}$$

With regard to the current state of the art in computational body counter calibration, available phantom series, and statistical analysis methods, extending the interpolation-based approach is a promising way to improve estimation of counting efficiencies due to the correlation with anthropometric parameters. The following changes can be made:

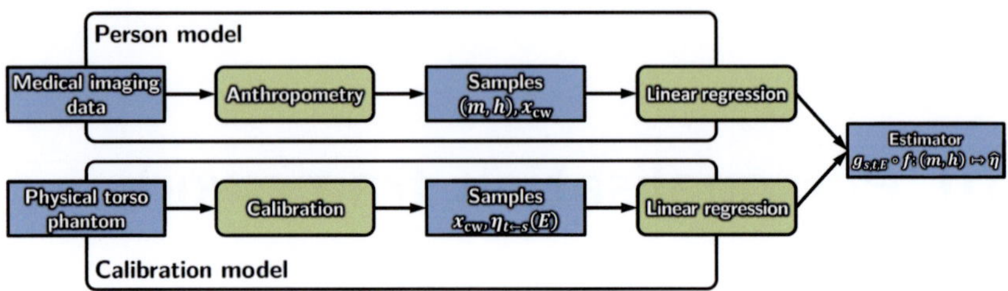

Figure 10.1: Workflow of interpolation-based approaches for personalisation of calibration factors using chest wall thickness x_{cw}. The estimator is a combination of linear estimators $g_{s,t,E} \colon (x_{cw}) \mapsto \hat{\eta}$ and $f \colon (m, h) \mapsto \hat{x}_{cw}$ based on two separate sets of sample data.

- Replace the separate base data sets for person and calibration model with a set of person-specific computational phantoms. This introduces dependency on the measurement setup into the person model and anatomical variation into the calibration model. Determination of chest wall thickness is unnecessary.

- Generalize the linear parametric models to a general non-parametric estimator f.

- Use photon energy E as an attribute of the estimator to provide estimates for radionuclides that have not been considered during model building and ensure consistency across the energy range.

- Extend body mass and height to a sequence of selected anthropometric parameter values $P = (p_1, \ldots, p_n)$ of a person or phantom.

- Move from a torso setup to any setup $t \leftarrow s$ with source $s \in S$ and detector $t \in T$.

These changes generalize the original model and improve its performance due to the added degrees of freedom and additional data.

$$\hat{\eta}_{t \leftarrow s}(E) = f_{s,t}(E, P) \tag{10.2}$$

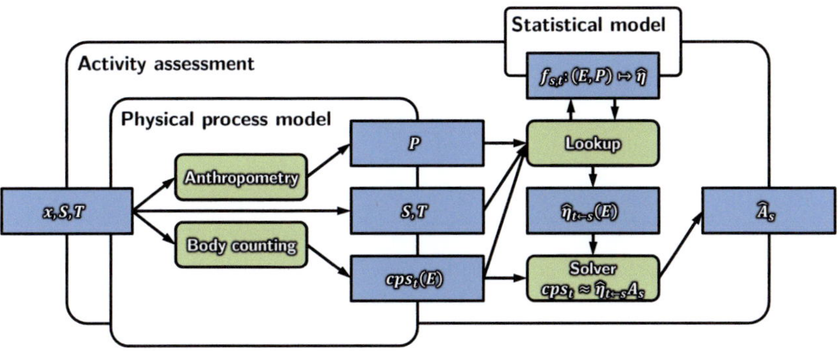

Figure 10.2: Workflow for activity assessment using body counting to determine count rates $cps_t(E)$ in specific measurement setups (S, T) and anthropometry to determine values P of anthropometric parameters. The physical process model specifies how to perform these tasks reproducibly. Querying these values in the statistical model results in estimated counting efficiencies $\hat{\eta}_{t\leftarrow s}(E)$ that are used to compute source activities.

10.2 Activity assessment

Activity assessment (figure 10.2) is the main process for application of the framework to actual measurements. It assumes that any kind of model $f_{s,t}\colon (E, P) \mapsto \hat{\eta}$ is available. So, the main concern is the measurement of these arguments. This procedure is actually a standard method in body counting, but requires formalization for integration with the framework. The main changes are formalization of detector positioning, consideration of additional anthropometric parameters, and provision of uncertainties for counting efficiency which must be propagated accordingly.

Given a person $x \in X$ that is supposed to be measured in setup $T \leftarrow S$, body counting is performed according to routine methods with the exception that detector positioning is performed according to a guideline that is specified in the physical process model for the body counter that was introduced with STEP. The measurement and subsequent evaluation results in count rates $cps_t(E)$ for the detectors $t \in T$ and photon energies E. In addition, a subset of anthropometric parameters P is selected depending on

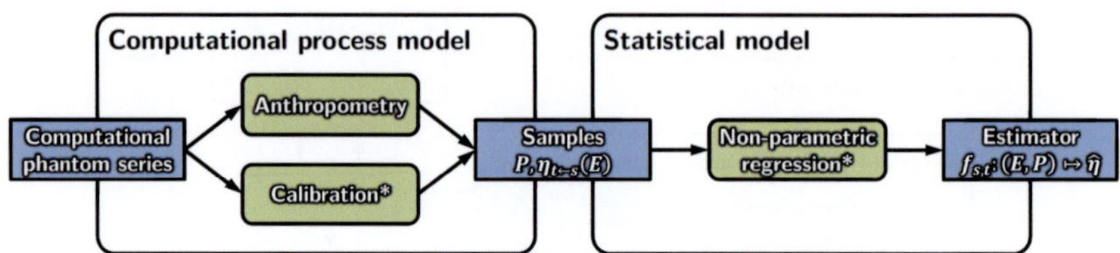

Figure 10.3: Workflow for model building applying computational anthropometry and body counter calibration to a series of phantoms (compare figure 10.1). Regression analysis is performed on the computed values P and $\eta_{t \leftarrow s}(E)$ to create an estimator $f_{s,t}\colon (E, P) \mapsto \hat{\eta}$ for counting efficiency. The calibration step is further described in figure 10.4 and the regression step in figure 13.2.

their scores in the statistical model and a possible cost to assess these. The measurement is also performed according to the physical process model.

The counting efficiency for each combination of s, t, E and P is determined by lookup or on-demand computation of the statistical model. The response is a set of estimated counting efficiencies $\hat{\eta}_{t \leftarrow s}(E)$ involving all detectors, sources and photon energies. Solving the system of equations $A_s = \hat{\eta}_{t \leftarrow s}(E)\, cps_t$ for all $s \in S$ and $t \in T$ results in estimated source activities $\hat{A}_s$.

10.3 Statistical model

The model building process (figure 10.3), which includes anthropometry and calibration of the detector system for a set of phantoms and the creation of a statistical model, is one of the main contributions of STEP. The measurement part is similar to activity assessment, but is being performed for computational phantoms and with respect to the computational process model, which is assisted by software to ensure reproducibility and improve computational efficiency.

Given a phantom $x \in X$ and measurement setup $t \leftarrow s$, calibration is performed for a set of predefined photon energies E covering the detector range. This results in a set of counting efficiencies $\eta_{t \leftarrow s}(E)$. In addition, anthropometry is performed, resulting in anthropometric parameter values

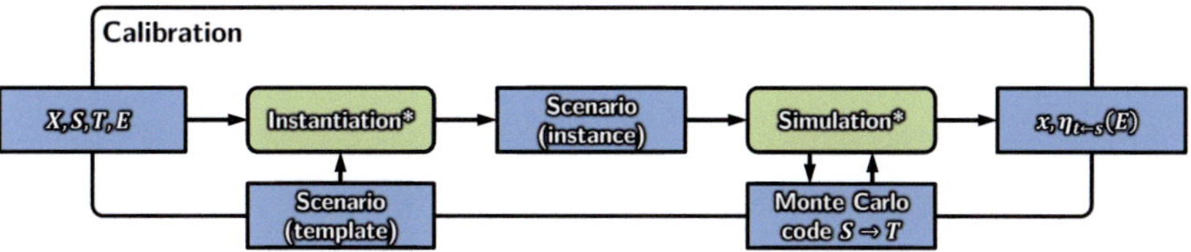

Figure 10.4: Workflow for computational body counter calibration. A template scenario of the measurement is instantiated with phantom series X. Radiation transport simulation is performed leading to a set of pulse-height spectra. The following analysis of the photo peaks results in counting efficiencies $\eta_{t \leftarrow s}(E)$ associated to each phantom $x \in X$. The instantiation step is further described in figure 12.1 and the simulation step in figure 12.4.

P. This process is repeated for each phantom of the series, and all values are collected in context of x.

The second part of the workflow is concerned with actually building the model $f_{s,t} \colon (E, P) \mapsto \hat{\eta}$. This is based on regression analysis for finding an optimal estimator that fits the samples well and also generalizes to unseen data.

10.4 Calibration

Computational body counter calibration (figure 10.4) is simplified by generating a template scenario including the detectors $t \in T$ and other parts of the body counter, a designated location for the phantoms X, and annotations for potential source locations $s \in S$. The template is instantiated to a full scenario by inserting the actual phantom, positioning the detectors and selecting source locations in the phantom according to the measurement setup $T \leftarrow S$ and the positioning guideline. The positioning strategy depends on the specific detector system.

Sources are homogeneously distributed in the associated structures of the phantom. Also, each photon energy in E is simulated separately as a mono-energetic source. And each potential source structure is chosen once as the source in a simulation, while recording the response of all associated detectors. This allows the quantification of crosstalk. After simulation with a radiation transport code, the photo peaks of the resulting pulse-height spectra are analysed and quantified as combinations of photon energy E and counting efficiency H for phantom x in measurement setup $T \leftarrow S$.

10.5 Anthropometry

The selection of anthropometric parameters and their computation is application- and implementation-specific. In general, parameters of interest are circumferences, diameters, weights, and skinfold thicknesses. The computational methods may use the orientation of prominent bones to identify palpable landmarks of the body and compute distances between points or circumferences in an intersecting plane defined by multiple points. This relies on basic algorithms for computational geometry (section 12.4).

10.6 Regression

The goal of the regression process is to build a model that estimates counting efficiency H based on assessed values P of anthropometric parameters for energies E and measurement setups $T \leftarrow S$. It is based on a machine learning approach for non-parametric regression using the computed results as observations of the unknown function $f_{s,t} \colon (E, P) \mapsto \eta_{t \leftarrow s}(E)$.

First, feature subset selection is performed on the full training data set. Then, given a selected subset, the data is projected to the new subset, an according model is created, and simultaneously applied to the unlabelled data resulting in labels consisting of mean and variance.

10.7 Implementation requirements

To ensure applicability of the method, it must be guaranteed that detector positioning and anthropometry are consistent among all phantoms and reproducible by technicians performing measurements on actual persons. The implemented methods must rely only on information available to them and be specified accordingly.

The described workflows of the STEP framework are relatively complex and their performance for a large number of phantoms is very time-consuming and repetitive. This may lead to mistakes when performing modelling tasks and introduce user bias when performing anthropometry and detector positioning due to different interpretation of the specification. The implementation of the framework should therefore automate as many processes as possible by integration in software tools based on according data models and providing a sense of modularity and abstraction.

The implementation of the STEP framework comprises three major components:

1. Development of an abstract and modular data model covering modelling, simulation and evaluation of radiation transport scenarios to facilitate data exchange and to reduce overall effort regarding data conversion and integration (chapter 11).

2. Development of a software tool implementing the data model while providing algorithms automatizing the individual processing steps for anthropometry and calibration based on the designated Monte Carlo code (chapter 12).

3. Implementation of the regression workflow with an existing statistical analysis tool and with regard to computational efficiency (chapter 13).

11 Data model

Modelling for radiation transport simulation is an essential task for prospective and retrospective evaluation of radiation transport scenarios. Many applications, such as calibration for whole and partial body counters or calculation of dose conversion coefficients for internal and external exposure, may include anatomically realistic computational phantoms with large amounts of data, geometrically complex detector models, or even require a series of simulations.

A key problem is handling large amounts of data from various sources that may be frequently updated and are provided in different file formats. They need to be compiled into corresponding syntax as input for the desired Monte Carlo code. The resulting simulation output typically requires additional post-processing.

Obviously, input formats of Monte Carlo codes cover all aspects of radiation transport such as geometric models, media properties, particle cross sections, and source and tally specifications. However, they are very specific to the code and only few modelling tools are available that are compatible with these formats. Various alternative data formats are already in use that cover some parts of the required data individually (e.g. phantom geometry, detector geometry, and radionuclide decay data). On the other hand, non-formal descriptions in textual or tabular form are common, for example, describing elemental composition of media or geometry-media mappings.

A holistic approach to a data model is needed that structures all data required for radiation transport and additional information required by the STEP framework. The developed model is inspired by the structure of MCNPX input, but not restricted to this specific code.

There are several requirements that the data model must fulfil:

- Modularity to support replacement of components
- Serialization to human- and machine-readable formats for data exchange
- Coverage of the data domains for body counting and anthropometry

The developed data model is called V2M SCHEMA (Pölz et al. 2013), because of its association with the software tool VOXEL2MCNP and its implementation in the markup language XML SCHEMA (XSD) (W3C 2004).

11.1 Modularity

Modularity of the data format is achieved by grouping semantically similar data as structures called *resources* and linking them via cross referencing. Resources may be organized in a local file system with each resource in a different file or multiple resources grouped into one file. References are uniform resource locators (URLs) identifying a file and exactly one resource in that file. They form a dependency relationship between the referencing and referenced resources. Through this approach, it is also possible to associate external source files from which data can be imported on demand.

11.2 Coverage

Monte Carlo codes for radiation transport generally require a basic set of data:

- Sources with probabilistic radiation emissions
- Tallies describing the computed quantity
- Objects with geometry and associated media
- Cross sections for radiation-matter interaction
- Simulation options regarding variance reduction and physics settings

In the current version, V2M SCHEMA defines eleven resource types (table 11.1): *Scenario, Equipment, Geometry, Binary, Materials, Elements, Source, Tally, Taxonomy, Simulation,* and *Results.* Each resource type has a basic set of attributes for identification via cross references and special subtypes for storing data. A schematic overview of the cross references is available in figure 11.1.

11.2.1 Scenario

A radiation transport scenario is defined as a collection of geometric objects with associated sources and tallies. Objects are instances of *Equipment* and

Name	Description
Binary	Compressed binary data for storing lattices.
Elements	A collection of chemical elements and their isotopes with associated mass attenuation coefficients and particle cross-sections.
Equipment	An abstract geometric object composed of segments with associated materials.
Geometry	A collection of cell volumes described by constructive solid geometry extended with repeated structures (lattices).
Materials	A collection of individual materials for describing density and elemental composition of geometric segments.
Results	Simulation results and derived properties specific to a scenario or any of its objects.
Scenario	A radiation transport scenario as a collection of objects arranged in space. Each object may be associated with sources and tallies.
Simulation	Settings regarding Monte Carlo simulations in general and specific to the code in use.
Source	A radiation source described as a discrete probability distribution of particle emission events.
Tally	The physical quantity to derive from the simulation and how to score each particle depending on location and energy.
Taxonomy	A hierarchical collection of terms describing structures and systems of equipment and materials semantically for identifying geometric segments and associating materials.

Table 11.1: Overview of resources defined in V2M SCHEMA (Pölz et al. 2013). Each resource is a self-contained entity describing a specific part important for radiation transport, but may depend on the availability of other resources.

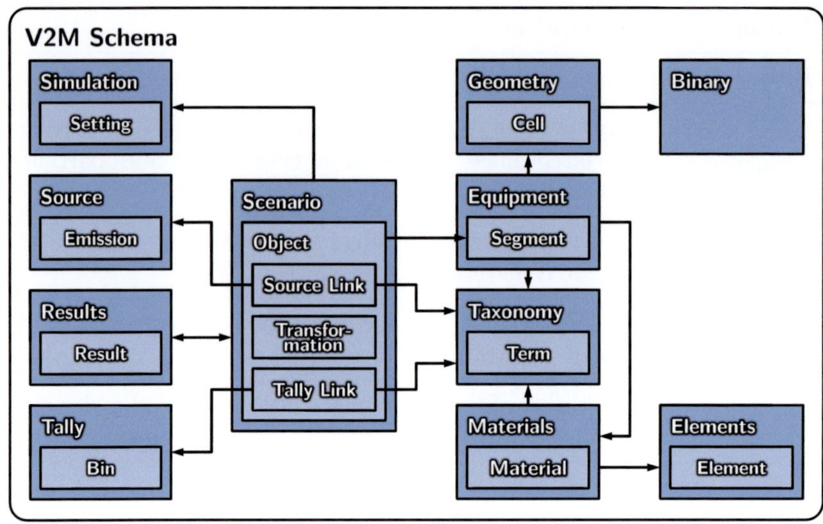

Figure 11.1: Overview of resources defined in V2M SCHEMA and their relationships (Pölz et al. 2013). Arrows indicate cross references in direction from the referencing object to the referenced object.

are arranged in space via affine transformation. Each object may contain source and tally links which define the location of the radioactive source and simulation tally and refer to *Source* and *Tally* definitions respectively with more detailed information. A location is defined via referenced terms of the equipment's *Taxonomys* with which the appropriate segment and cell volume or its surface can be identified. Additionally, settings for calculating the scenario with a Monte Carlo code are defined in an associated *Simulation*, and simulation results are stored in form of *Results*.

11.2.2 Equipment

An equipment is an abstract geometric object composed of distinct segments with associated materials. It references a *Geometry* which defines these segments geometrically, and *Materials* which describe their material properties. A segment-term mapping provides semantics for the segments in context of the *Taxonomy*. A segment-material mapping defines the direct association of materials.

11.2.3 Geometry

Geometry is described as composition of complex shapes with geometrical primitives in form of constructive solid geometry (section 5.2). It is essentially a collection of cells which are of the following three types:

Simple cells are volumes defined by primitive surfaces such as sphere, box, plane, cylinder, cone, or torus which may be parameterized and transformed via affine transformation to create non-unit primitives.

Complex cells are combinations of multiple simple or other complex cells with a set operator such as union, intersection, difference, or complement. They allow the creation of non-primitive geometries.

Repeated cells are special cells for implementing voxel phantoms. Similar to MCNPX, they are repeated structures of a certain primitive (box or hexagonal prism) arranged in a three-dimensional lattice that is embedded in a simple cell. The actual lattice data is stored in a referenced *Binary*. The repeated cell only specifies the interpretation of the data by defining a template cell and lattice dimensions.

Additionally, any top-level cell of the cell hierarchy must define a segment identifier which is in turn referenced by the related equipment. Other cells, i.e. those used to form a complex, are implicit and do not require materials.

11.2.4 Binary

Repated cells can be quite large depending on voxel resolution. They are modelled as a list of integer or decimal numbers in a specific order. For example, a three-dimensional lattice with segment identifiers ranging from 0 to 255 is stored as a linear 8-bit integer array. This is sufficient for most phantoms. Larger ranges are possible by increasing the element size.

11.2.5 Materials

Similar materials or those required by the same equipment are grouped into a *Materials* resource. Each material has name, density and elemental composition. The elemental composition defines mass or atomic fractions for each element contained. Associated *Elements* further describe the material components on the elemental level. For the segment-material mapping, each material is also associated with a set of terms from possibly multiple taxonomies that are referenced.

11.2.6 Elements

Properties of chemical elements or their isotopes are stored in an *Elements* resource. Each element is described by basic properties, e.g. atomic number, mass number, chemical symbol, name, and density. Mass attenuation coefficients and particle cross-sections can be associated for performing radiation transport calculations, for example for calculating linear attenuation coefficients of photons through matter.

11.2.7 Source

Radiation sources are described as discrete probability distributions of particle emission events (e.g. during radioactive decay). Each event consists of energy, intensity (probability), and particle type. Continuous spectra (e.g. from beta decay) can only be represented approximately using discretization. Also, additional information regarding radioactive decay can be given. For example, decay products for modelling dynamic sources with nuclide fractions changing over time.

11.2.8 Tally

A tally contains information indicating the physical quantity to derive from the simulation, such as particle flux, current, deposited energy, or pulses created in a detector, and how to score it with respect to space and energy. For calculating dose conversion coefficients, for example, the total deposited energy per tally volume is needed. For counting efficiency calibration, the total number and energy of pulses created in the tally volume (active detector crystal) by interacting particles is scored. Energy discretization can be added by defining an amount of bins and an energy range. Information about energy resolution of the detection device can be stored by setting parameters of a standard model (equation 3.1) describing energy dependency.

11.2.9 Taxonomy

Taxonomies are used as hierarchical collections of terms describing structures and systems of phantoms, detectors, other equipment, and even materials. Geometry segments and materials can be registered in one or multiple taxonomies to associate them with semantics. Based on these associated terms matchmaking algorithms can be applied to suggest an initial mapping between the defined segments and materials.

11.2.10 Simulation

Monte Carlo codes require certain settings specific to the problem they are simulating. These settings are either related to the Monte Carlo method in general or specific to the actual code. The following settings can be stored:

- Problem termination conditions such as maximum number of particle histories to be computed, maximum computer time, or maximum relative standard deviation for each tally.

- Various options for variance reduction, such as a filter for energy and intensity range for the source to omit negligible emissions or a selection of which secondary particles to transport.

- A threshold setting the minimum sampling efficiency of position sampling for source particles. A higher threshold may reduce memory usage, but may also increase computer time.

11.2.11 Results

Simulation results are collections of attribute values for storing tally data and derived properties specific to a scenario or any of its objects. Each item may have attribute name (physical quantity), source identifier, tally identifier (cell or surface), energy (or energy range), value, and relative error (uncertainty of the value as estimated relative standard deviation).

Source and tally identify corresponding resources in the corresponding scenario. This is sufficient to model simulation results as well as geometric properties of the objects, e.g. anthropometric parameters and detector properties. Unnecessary values can be omitted. For instance, most anthropometric parameters do not depend on source, tally and energy. An exception to this is chest wall thickness.

11.3 Serialization

A modular data model was created to facilitate the description of all components. XML SCHEMA (XSD) (W3C 2004) was chosen as an abstract modelling language to describe the data model. It allows the creation of complex, but human-readable and easily modifiable data formats. Instances of the created schema, called V2M SCHEMA, are stored as `.v2m` files and can be validated against their specification and parsed with freely available software libraries, such as QTXML (Qt Project and Digia 2013), LIBXML2 (Veillard 2012) and JAXB (Java Community 2012).

12 Software implementation

The software tool VOXEL2MCNP (Hegenbart et al. 2012; Pölz et al. 2013) for computational body counter calibration is under development at KIT. Starting as a command line tool for converting voxel models into MCNPX syntax in 2008, many features were added, including two- and three-dimensional visualization, a model of the body counter with phoswich detectors at IVM, and methods for modifying voxel models. It is used by staff members and students for radiation transport simulation tasks for a variety of projects.

For integration with the STEP framework, a major redesign of the software was necessary to provide the level of abstraction, modularity, and efficiency required for performing sensitivity analyses with a large series of computational phantoms. The goal was to support the anthropometry and calibration processes including modelling, simulation management, and measurement evaluation.

12.1 Application structure

VOXEL2MCNP has been redesigned from scratch maintaining and extending functionality of the original software while integrating the developed V2M SCHEMA. The software is being written in C++ using the cross-platform application and user interface framework QT (Qt Project and Digia 2013) and the graphics library OPENGL (Shreiner 1999). VOXEL2MCNP is platform independent and compilations for MICROSOFT WINDOWS, MAC OS X and several LINUX distributions have been tested.

The application consists of QT widgets facilitating user interaction, libraries offering data models and computational algorithms, plug-ins providing methods for file import and export, and a core. The application core is responsible for plug-in management and file handling, provides a basic graphical component for integration of widgets, and a data manager and event-based notification system for those widgets.

Libraries provide stand-alone functionalities and data types, such as computational algorithms for surface triangulation for visualization of scenarios, adaptations of several data formats used by plug-ins, and so on. As described in chapter 5, many popular modelling tools exist. The goal of VOXEL2MCNP is not to mimic the functionalities of those tools, but to provide means to access models created with these through the abstract interface of V2M SCHEMA. An object-oriented adaptation of the developed V2M SCHEMA is also implemented as a library providing additional data management functionality.

Supported file formats are implemented as plug-ins which conform to a common interface offering file import and export. They are discovered automatically on application startup.

Widgets provide functionality requiring user interaction. The basic functionalities of the original VOXEL2MCNP have been abstracted to widgets and extended. This includes a transaxial view displaying voxel models in slices, a perspective view visualizing an interactive scenario including all objects with functionality for automatic or interactive detector positioning, an editor widget representing data in editable forms and tables, a properties widget giving information about anthropometric parameters and segment volumes and masses.

12.2 Data import and export

Several plug-ins (table 12.1) have been created to perform data conversion in form of file import and export. When a file in V2M SCHEMA format is loaded, all referenced files are registered with the file handler. When a reference needs to be resolved because of a data request, a file plug-in with proper capabilities is automatically selected and executed.

12.2.1 V2M Schema

The XML version of V2M SCHEMA is the native format of VOXEL2MCNP. Each resource is serializable to an according XML element and a .v2m file can be validated against the schema file. Data compression is applied in case of binary resources.

File format	Resource	Import	Export
V2M SCHEMA	Any	Yes	Yes
IMAGEJ	Geometry	Yes	Yes
SIMPLEGEO	Geometry	Yes	Yes
ENSDF	Emissions	Yes	(No)
MCTAL	Results	Yes	(No)
MCNPX	Scenario	No	Yes

Table 12.1: Overview of file formats supported by VOXEL2MCNP and their associated resources in V2M SCHEMA.

12.2.2 ImageJ

IMAGEJ (Ferreira and Rasband 2012) is an image processing program developed at the National Institutes of Health (NIH). It also supports image series, which is the native representation of tomographic imaging data used for voxel models. Such a three-dimensional lattice is stored as a linear representation in a binary file with an external header file describing the structure of the binary.

The according plug-in designed for VOXEL2MCNP converts this format to V2M binary. The structural information is stored in an associated geometry with a repeated structure cell based on a cuboid surface.

12.2.3 SimpleGeo

SIMPLEGEO (Theis et al. 2006) is a three-dimensional modelling tool developed at CERN. It was specifically created to unify the various geometry modelling processes and syntaxes of radiation transport codes. A key feature of SIMPLEGEO is that it stores data in a (custom) constructive solid geometry (CSG) format instead of a boundary representation (B-rep) format (Mortenson 1985). This is of importance, because a conversion from constructive solid geometry to boundary representation, which is basically an approximation of volume boundaries with planar faces or free-form surfaces, results in information loss.

The according plug-in designed for VOXEL2MCNP converts this format to V2M geometry. SIMPLEGEO uses a tree with parametric surfaces at the leaves and set operators at the inner nodes. It also associates surfaces with sets of shading parameters. Each surface is converted to one of the basic surfaces of V2M SCHEMA and the parameterization is stored in the associated transformation matrix. Inner nodes are converted to complex cells. The identifier associating the shading parameters is used as associated segment identifier for each cell.

12.2.4 ENSDF

ENSDF (Tuli 2001) is maintained by the National Nuclear Data Center (NNDC) at Brookhaven National Laboratory (BNL). It stores nuclear structure and decay data for radionuclides in form of data records similar to MCNPX. Each record has a type (e.g. gamma emission) and specifies an according set of parameters.

The according plug-in designed for VOXEL2MCNP converts this format to V2M emissions. All records that signify an emission are identified and the appropriate information extracted to convert to a V2M emission.

12.2.5 MCTAL

MCTAL (Pelowitz 2007) is the formal output format of MCNPX. It is a listing of tally values arranged by tally binning. Each tally is represented by identifier, binning, value for each bin with relative error, and the tally fluctuation chart for statistical tests.

The according plug-in designed for VOXEL2MCNP converts this format to V2M results. The results are automatically associated with the scenario that was used to create this output and sources and tallies are identified based on tokens that are placed by the MCNPX export plug-in.

12.2.6 MCNPX

MCNPX was designated as the primary radiation transport code for performing simulations with VOXEL2MCNP because of the preexisting experience of coworkers with the code and the simple structure of its input files, which makes it easy to perform code generation from V2M SCHEMA. MCNPX input files (Pelowitz 2007) are composed of surfaces, cells, transformations, materials, sources, tallies, and settings with regard to radiation transport and variance reduction.

MCNPX file export is achieved by creating MCNPX cards for all objects in a V2M scenario. Each MCNPX card has associated resource elements, e.g. surface and cell cards are created from V2M geometry, material cards are created from V2M materials and so on. For managing the various numeric identifiers MCNPX uses, namespaces have been introduced handling surface, cell, transformation, material, source distribution and tally identifiers.

A simulation series for a phantom with changing source emissions and source locations, can be specified by a single V2M scenario. The export plug-in can create a set of MCNPX input files that perform each of the combinations of emissions and locations while reducing redundancy by storing duplicate parts in common files that are referenced.

12.3 Scenario modelling

The main intent of VOXEL2MCNP is to support modelling for radiation transport in body counting scenarios. The idea is that as many steps as possible are undertaken using dedicated tools whose files are then imported, and then integrated into a V2M scenario (figure 12.1). Beginning from a scenario template, each step integrates or replaces a model and performs further adjustments. Several user interfaces which allow visualization, editing, computation, and validation are provided to finalize a scenario model.

12.3.1 Resource editor

The full V2M SCHEMA has an additional representation through QT widgets. These are graphical components which allow user interaction and editing (figure 12.2). Any data that may not be imported via plug-ins can be imported manually by creating an according .v2m file or using the editor.

12.3.2 Material annotation

To perform radiation transport, Monte Carlo codes require density and material composition of all media involved, either defined as mass fractions or atomic fractions of elements and their respective isotopes. Some definitions may be intended for special applications requiring a certain level of detail from the geometric model, or assume different segment semantics. With

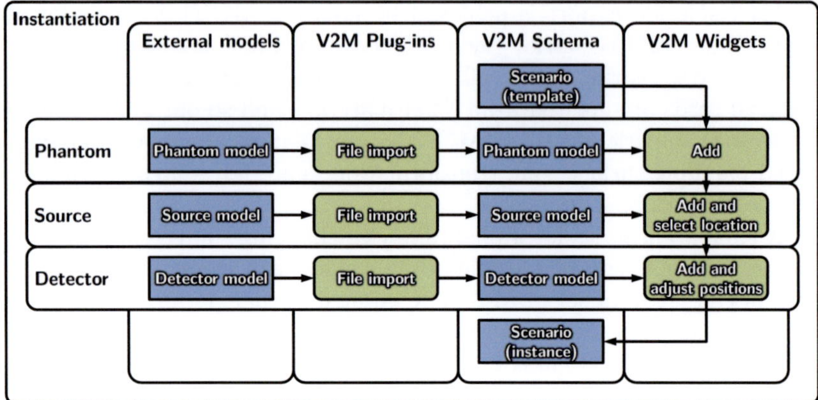

Figure 12.1: Workflow for instantiation of a template scenario (Pölz et al. 2013). Individual models are created with dedicated modelling tools, imported into VOXEL2MCNP and combined to a full body counting scenario.

regard to computational phantoms based on medical imaging data for example, a medium named "lung tissue" may specifically refer to bronchial and alveolar tissue, but it may also include different portions of blood and air depending on imaging and segmentation techniques applied.

Aside from technical issues, phantom developers may simply group similar structures differently dependent on application or personal preference. These differences pose a problem when assigning materials to geometry segments; allowing only fixed segment-material mappings limits the intended modularity of the data model. Also, exchanging material libraries for certain objects can be useful, for example, for performing sensitivity analyses or for switching to alternative versions. This is not possible without re-assigning each segment to its corresponding material of the new library, because there is no sense of semantics associated to the individual material names, which can be processed by a computer.

To address these issues and preserve modularity of the data model, a new approach using controlled vocabularies was implemented. A taxonomy (National Information Standards Organization 2005) is a structured set of abstract terms where each term is associated with a detailed description. Here, taxonomies are used as a common vocabulary to associate segments

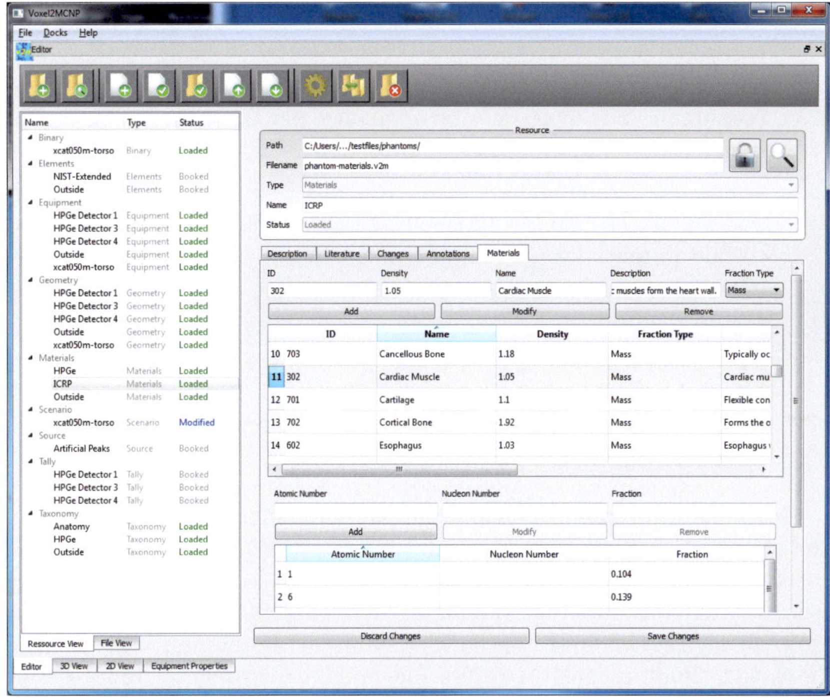

Figure 12.2: Editor module of VOXEL2MCNP showing the associated resources of a loaded body counting scenario (*left*) and details of the selected resource (*right*).

and materials with semantics. The individual terms are structured in a hierarchy with a binary relation to provide different abstraction layers, i.e. there is a single root term at the top of the hierarchy and each term may specialize into several disjoint categories. When importing a new equipment or a new material library segment-material mapping is performed with the following method:

1. If the equipment or material library is not covered by the existing taxonomies, one or more taxonomies are designed to describe the domain of objects in question by selecting and categorizing terms related to this domain.

2. For each equipment, all segments of the associated geometry are registered in the corresponding taxonomies by selecting the most specific of the matching terms available. This mapping can be quite time-consuming, but must only be done once per taxonomy. This is also done for all material libraries of interest using the same taxonomies.

3. A heuristic matchmaking algorithm based on similarity of segment and material semantics (defined by the associated terms) is applied to find potentially matching segment-material pairs depending on what is actually required by the equipment and provided by the material library.

4. The initial mapping is revised and fixed in the equipment definition for each of the selected material libraries. The amount of work required for this step can vary significantly and largely depends on the effort put into the first three steps.

In general, the use of multiple simple taxonomies is encouraged over the use of a single complex taxonomy, because they may provide different categorizations of the same terms or allow the combination of terms from related but different domains. For example, taxonomies about (superficial) regions and anatomic systems of the human body can be employed simultaneously to describe human anatomy in more detail. For computational phantoms, existing taxonomies for anatomic system classification and histologic classification can be adapted for use. In case of detectors and other objects, very simple taxonomies can be designed, if there is no imminent need for exchanging material libraries.

VOXEL2MCNP provides a transaxial and perspective view to visually identify segments of an equipment and an editor to associate terms. They can be used in combination to perform annotation or revision while comparing the appearance of the segments and the semantics of the associated terms.

For a material to be applicable to a segment, it is necessary that each material term has a corresponding segment term in the taxonomy that is equal or more specific, so that in principal the most general tissue is always applicable. The matchmaking algorithm is based on a distance metric to compute a score for each material and rank them accordingly. The optimal material for any segment is the one which minimizes this score and therefore the distance of the corresponding terms over all taxonomies.

12.3.3 Scenario visualization

VOXEL2MCNP stores its geometry equivalent to the data model in the form of constructive solid geometry, which basically describes the hierarchical construction of volumes using geometric primitives and set operators. For visualization, these volumes are first converted to boundary representation via surface triangulation and then rendered with OPENGL.

There are several methods well-known in computer graphics to perform surface triangulation, such as marching cubes (Lorensen and Cline 1987), dual contouring (Ju et al. 2002), or sphere tracing (Hart 1996). The first two methods have been implemented in VOXEL2MCNP using extensions of the basic algorithms (Nielson and Hamann 1991; Schaefer, Ju and Warren 2007). The implementation of sphere tracing is still work in progress, but promises to further improve computational efficiency and visual quality.

12.3.4 Detector positioning

Modelling body counting scenarios requires accurate detector positions and orientations relative to the phantom. To support the task of detector positioning, VOXEL2MCNP provides an abstract module that can hold implementations of direct and inverse detector kinematic specific to a body counter.

The direct detector kinematic allows setting each free parameter of the detector mechanics while the detector position and orientation are automatically derived. This is useful for measurement reconstruction, where such parameters are recorded.

The inverse detector kinematic allows setting detector position and orientation, and the elongations and angles of the mechanics are automatically derived. This is useful for measurement planning, where the detectors are assigned to a target organ and positioned based on anatomical landmarks in that region or the organ itself (figure 12.3).

Additionally, if position of the detector and the target are known, the distance from the front centre of the entrance window to the skin of the body is given by intersecting the corresponding line segment with the phantom. Adjusting the detector position changes the distance accordingly.

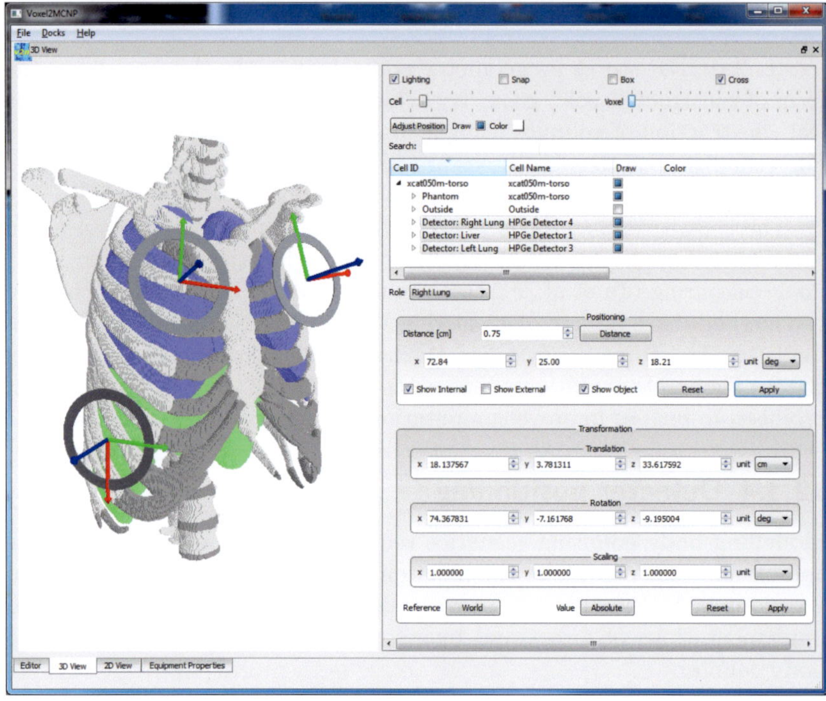

Figure 12.3: Perspective view and positioning module of VOXEL2MCNP for a loaded body counting scenario. Visible are parts of the phantom and locations of the three selected detectors (*left*), and controls for visualization and positioning (*right*).

12.4 Anthropometry

Body counting is sensitive to individual anatomy for many scenarios involving low-energy photons. One way to quantify anatomy is to describe it with geometric features, e.g. thicknesses, distances, lengths, breadths, circumferences, volumes, and masses. Such measurements are usually used in health examination studies, e.g. the National Health and Nutrition Examination Survey (NHANES) (Department of Health and Human Services and National Center for Health Statistics 1996) of the Centers for Disease Control and Prevention (CDC), and the clothing industry (European Committee for Standardization 2001). Acquisition of these parameters is a precondition for comparing phantoms and persons.

The VOXEL2MCNP anthropometry module provides a collection of methods to compute anthropometric parameters for computational (voxel) phantoms.

12.4.1 General parameters

In general, the module uses the annotated segments of phantoms to identify landmarks of the human body (similar to heuristic detector positioning). This method largely depends on the availability of the correct segments in the geometry of the phantom. If they are not available, heuristics based on anatomical background knowledge can be used to estimate the correct measurement positions.

The mass of segments can be computed from their geometrical volumes and associated material densities. Major structures can be used to quantify the location and orientation of body parts and anatomic landmarks. The orientation of those parts can be analysed using principal component analysis (Abdi and Williams 2010) on the set of voxels.

For computing circumferences, a cutting plane can be constructed with a certain position and orientation intersecting the phantom. A circumference is then computed from the convex hull (Graham 1972) of the intersection. Several methods were tested to improve results due to aliasing effects on voxels. Circumferences are usually evaluated at multiple locations along the main axis of the cutting plane to determine a minimum or maximum in that region depending on the quantity.

12.4.2 Average photon transmission

An anthropometric parameter that is very sensitive to body counting for lung measurements is effective chest wall thickness. It basically describes the average thickness of the absorber material between source organ and detector. Measurement and calculation of this quantity must take detector position, crystal dimension, and tissue composition into account. Physical measurement of chest wall thickness has already been performed by Gün (2010) with equipment available at KIT in collaboration with the medical centre and an according algorithm was implemented for voxel phantoms in VOXEL2MCNP.

To provide this parameter for measurement of all source organs and structures, the algorithm was redesigned to compute the average transmission of narrow photon beams between body surface and source depending on detector position and orientation. Normalization to reference muscle tissue resulting in an effective tissue thickness is omitted and instead photon transmission is used to provide an abstract measure that is linearly related to counting efficiency.

1. Produce random points on the detector entrance window and create according rays parallel to its surface normal.

2. Intersect each ray with the section of the phantom between detector and source. Discard rays missing the source.

3. Compute the relative track length for each ray in all media of the tissue.

4. Determine complex mass attenuation coefficients using the relative track lengths as weights (equation 6.2).

5. Average the transmission of each ray (equation 6.3), but omit normalization.

Elemental mass attenuation coefficients are derived from data (Hubbell and Seltzer 2004) using log-log cubic-spline interpolation over energy as proposed by Berger et al. (2010).

12.5 Simulation and evaluation

For simulation of a created scenario, VOXEL2MCNP exports the scenario to MCNPX syntax after any simulation-specific preprocessing (e.g. simplification). If energy resolution is known *a priori*, simulations can be run with the built-in option for Gaussian energy broadening (GEB) of MCNPX. Alternatively, GEB can be applied *a posteriori* according to the MCNPX model (equation 3.1) using the parameters describing FWHM stored in the detector equipment. The results are imported afterwards. Post-processing covers the evaluation of simulated pulse-height spectra by offering background subtraction with a linear or step model (section 3.3).

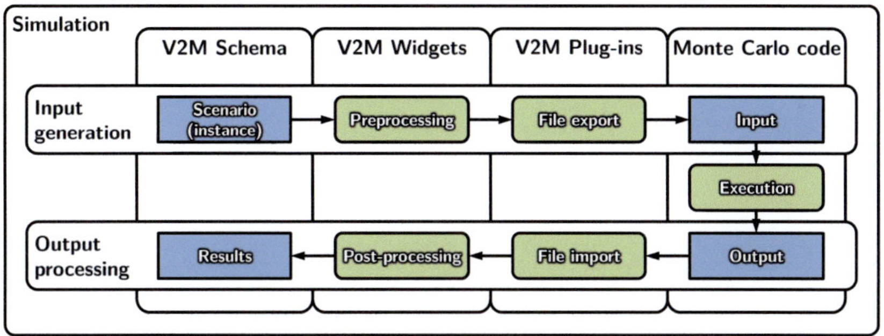

Figure 12.4: Workflow for simulation of a body counting scenario including input generation from a scenario instance, simulation execution with a dedicated Monte Carlo code, and processing of the simulation output (Pölz et al. 2013).

Adding several methods for energy interpolation (Gray and Ahmad 1985; Kramer 2007) of counting efficiencies is planned, but currently performed with the NLS2 package (Grothendieck 2010) of the statistical analysis language and environment R (The R Core Team 2013), which uses a Newton-type gradient descent.

13 Statistical analysis method

The workflows in this chapter are based on a specific graphical notation. An overview of the graphic components is given in figure 13.1.

The goal of the regression process (figure 13.2) is to build a model that predicts counting efficiency based on assessed values of anthropometric parameters for different energies and measurement setups. It was realised with a machine learning approach for function regression using the computed results as observations of the unknown target function predicting counting efficiency from anthropometric parameters.

First, feature subset selection is performed on the full training data set. Then, given a selected subset, the data is projected to the new subset, an according model is created, and simultaneously applied to the unlabelled data resulting in labels consisting of mean and variance.

13.1 Subset prediction

A resampling method is applied to the data creating a set of folds. Each fold is used for training an estimator. The ensemble of estimators is combined by averaging, resulting in mean and variance (figure 13.3).

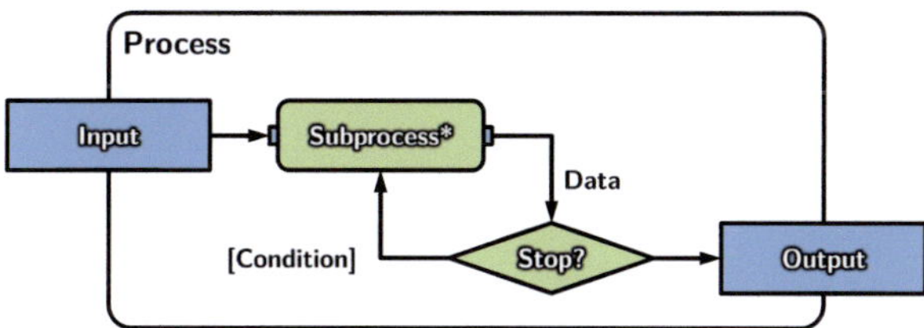

Figure 13.1: Graphic components used in workflow diagrams. An asterisk indicates that a detailed version of the process is presented in another diagram.

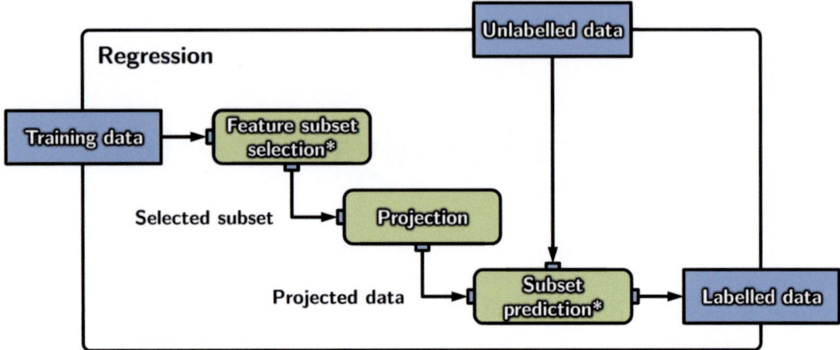

Figure 13.2: Workflow for regression applying feature subset selection in a preprocessing step on training data, then training the estimator on the projected training data, and finally applying it to unlabelled data.

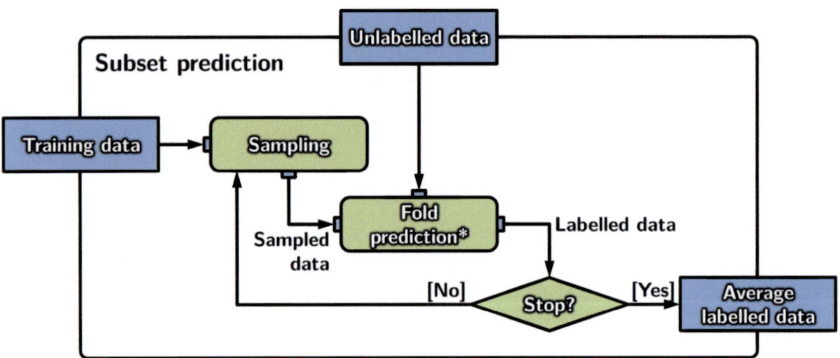

Figure 13.3: Workflow for subset prediction using resampling to create an ensemble of estimators which are then averaged.

An appropriate resampling method (section 8.2) may be selected depending on the data. Bootstrapping is used by default, which makes this process an implementation of bagging (section 8.3).

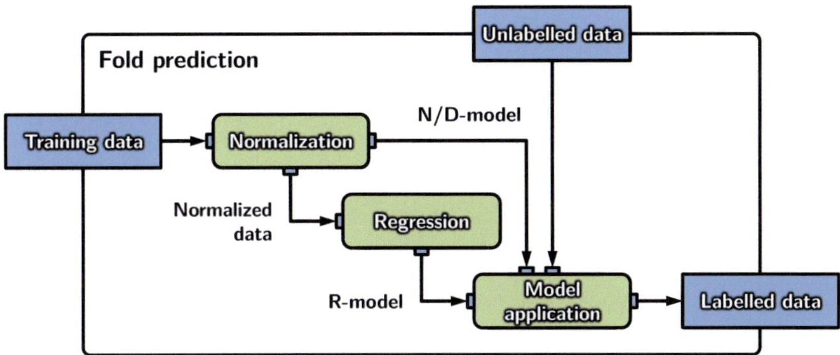

Figure 13.4: Workflow for fold prediction building a regression model from normalized data and applying it to unlabelled data in the order of normalization (N), prediction (R), and denormalization (D).

13.2 Fold prediction

Data introduces bias in a regression model due to different domains of the parameters. Therefore, as a preprocessing step, the training data is normalized using Z-transformation to remove its bias and normalize its variance. The normalized data is then fed to a regression method that creates an implicit model of the estimator, which is then applied to the unlabelled data. The unlabelled data is also being normalized before model application and denormalized afterwards (figure 13.4).

Any regression method (chapter 7) may be used for model building. Local polynomial regression is chosen by default.

13.3 Feature subset selection

To ensure an accurate, but also general model, feature subset selection (figure 13.5) is necessary to reduce the attribute domain. It follows the concept of minimum redundancy maximum relevance (section 8.5) using a preselection strategy, a search method, and a performance measure.

As a first step, a heuristic preselection may be performed using a correlation-based filter method or by manually removing irrelevant

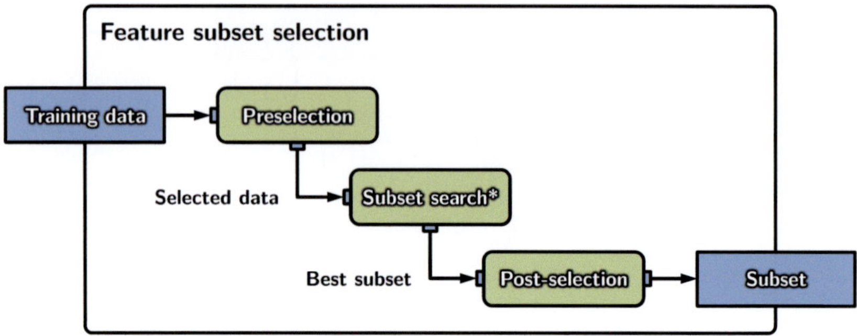

Figure 13.5: Workflow for feature subset selection. Optionally, filters may be applied to heuristically identify relevant features before performing an automated subset search. Afterwards, a manual examination of the evaluated subsets may be necessary.

features depending on the measurement setup. This is advised only if computational efficiency is a problem. The actual subset search is capable of finding appropriate feature subsets by itself.

The subset search minimizes a measure derived from the performance of the training data. After termination, the best subset is selected. However, any number of subsets and their scores may be extracted.

13.4 Subset search

Subset search (figure 13.6) is the process of (efficiently) moving through the space of feature subsets, evaluating their performances, and finding an optimum.

The selection of a search strategy (section 8.4) depends on the total number of features and the required computational efficiency. The general approach is to start with a maximum subset size of one which iteratively increases. Each subset is evaluated by computing a score according to a specific measure, which is used to guide the feature subset space search. A simple and guaranteed optimal method is exhaustive search. If this is not an option, a genetic algorithm (Goldberg 1989) may be selected.

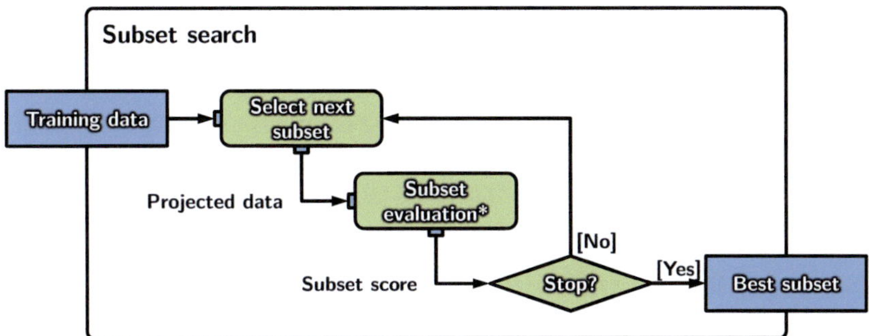

Figure 13.6: Workflow for subset search using a heuristic search strategy and a performance measure to evaluate each feature subset.

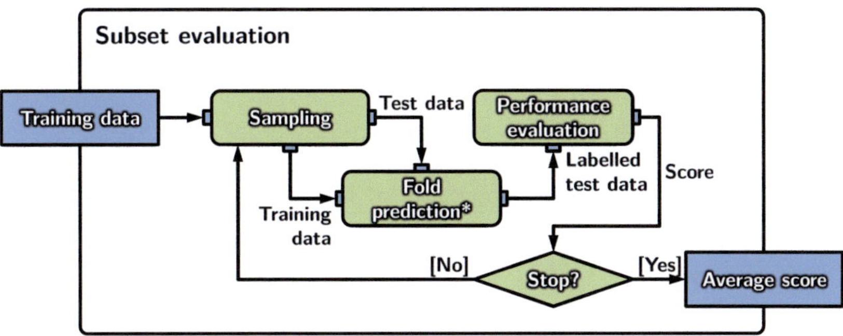

Figure 13.7: Workflow for subset evaluation using the wrapper method with resampling to get robust estimates for the subset scores.

13.5 Subset evaluation

The task of the subset evaluation process (figure 13.7) is to evaluate the performance of a feature subset given the training data. The wrapper approach is applied to do this effectively. This means that instead of using a filter, the same regression method that is used to build the final estimator is also used, for each feature subset.

The training data is resampled (e.g. cross sampling or bootstrapping) into different folds for training and testing. The regression method is applied to the new training data, and the resulting model is applied on the testing data whose performance is evaluated by a performance measure (section 8.1) for regression. This is repeated a certain number of times and the performances are averaged to a subset score.

13.6 Implementation

RAPIDMINER (Mierswa et al. 2006) is a popular open source data mining tool. It is under development since 2001 and regularly updated. It provides a broad range of methods used in machine learning, statistical analysis, and data mining. Those methods are available as parametric building blocks that are connected via data flows. Several example applications of RAPIDMINER for feature selection are available in Schowe (2010).

The implementation of the whole STEP regression workflow has been done with RAPIDMINER in version 5.3. The described processes were replaced with predefined customizable building blocks. One major drawback is that the kernel modules (e.g. local polynomial and support vector regression) do not provide uncertainties for their generated estimators. This problem could not be solved directly inside the tool.

Part III

Application

14 IVM body counter

The new body counting system at the KIT *in vivo* measurement laboratory (IVM) (Marzocchi 2011) includes four high-purity germanium detectors (HPGe) with thermoelectric cooling. It supersedes the old system based on two NaI(Tl)/CsI(Tl)/NaI phoswich detectors (Hegenbart 2009). The new system is an improvement in terms of energy resolution and spatial sensitivity. The lower detector size allows an increase in degrees of freedom regarding detector positioning.

The STEP framework has been applied to the IVM body counter first. The system was modelled with VOXEL2MCNP to perform computational body counter calibration with regard to current equipment and procedures in use. Computational phantoms were annotated to allow identification of certain structures for assigning sources, detector positioning, and calculation of anthropometric parameters.

Various applications with regard to body counting using specific computational phantoms are described in the following chapters:

- Measurements of the JAERI phantom were reconstructed and simulated to evaluate the validity of the computational approach (chapter 15).

- Four phantoms implementing the ICRP reference man specification were compared to check if organ masses are a sufficient constraint in phantom development (chapter 16).

- As the primary application of this work, the STEP framework was applied to the XCAT series to quantify the impact of interindividual anatomical variation (chapter 17).

- Inhomogeneous source distributions were created with a perforated lung set of the LLNL phantom and measured to estimate the uncertainty introduced by the general assumption of homogeneous distributions (chapter 18).

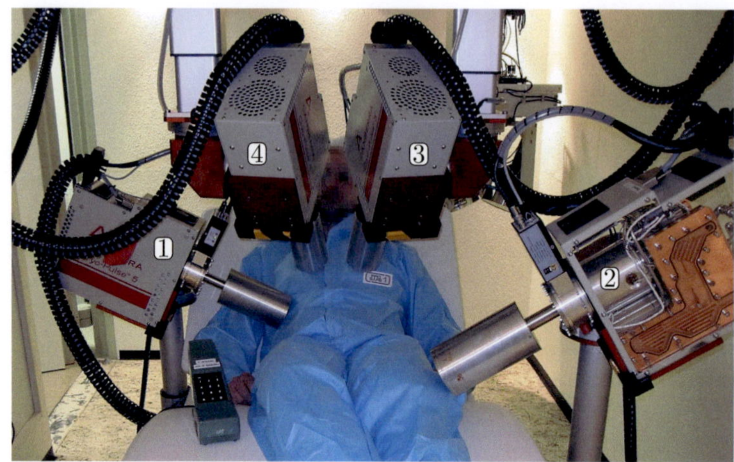

Figure 14.1: Measurement chamber with four freely arrangeable HPGe detectors and an adjustable stretcher. The detectors are arranged for activity assessment in (1) liver, (2) bone, (3) left lung and (4) right lung.

- A study was performed to quantify the effect of respiratory motion using four-dimensional computed tomography data as an example of intraindividual anatomical variation (chapter 19).

14.1 Equipment

The IVM body counter includes a measurement chamber with four freely arrangeable detectors, an adjustable stretcher, and a device for tracking and changing detector positions (figure 14.1).

14.1.1 Detectors

The detectors are extended range coaxial (XtRa) germanium detectors by Canberra (2013). They have an efficiency of about 80 % relative to reference NaI(Tl) detectors and a maximum operational energy of 2.048 MeV. The energy range is divided into 8192 channels with 0.25 keV

Figure 14.2: XtRa HPGe detector with carbon entrance window *(front)*, aluminium casing around the germanium crystal, preamplifier *(top centre)*, and thermoelectric cooling system *(back)*.

channel width. The energy resolution (FWHM) is about 0.8–2.5 keV across the operational range.

The front casing (figures 14.2 and 14.3) contains a copper frame holding the germanium crystal and a thin carbon composite entrance window with a diameter of 7.5 cm. The pulses are preamplified and sent to an external processing unit. The back casing contains the electric cooling system with a fan-based heat exchanger.

Detector geometries have been modelled in MCNPX syntax and are continuously optimized with respect to crystal dimensions, front dead layer and copper frame using point source measurements, radiography of the detector head, and parametric simulations (Elanique et al. 2012; Marzocchi, Breustedt and Urban 2010).

For use with VOXEL2MCNP, the geometries were modelled with SIMPLEGEO by Laubersheimer (2012). Equipment files (listing A.2) have been created for each of the four detectors containing references to the geometries stored in the according SIMPLEGEO files (figure 14.3). Data conversion is automatically performed by VOXEL2MCNP on file import

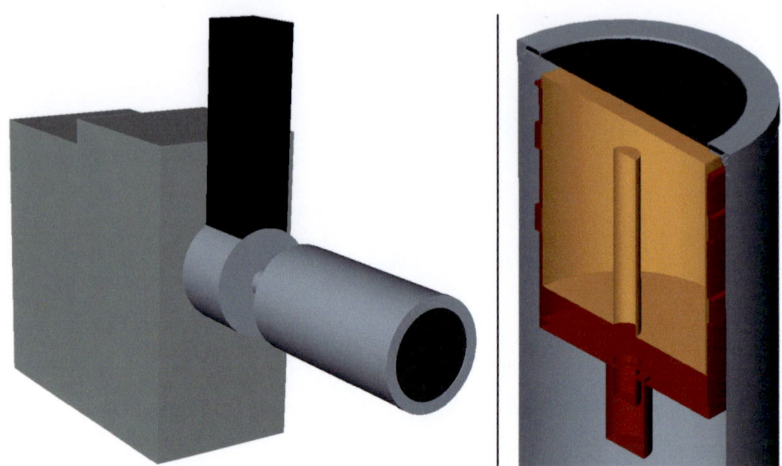

Figure 14.3: SIMPLEGEO model of one of the four HPGe detectors. *Left:* The detector crystal is located inside the cylindrical frame in the front case. The preamplifier and case of the cooling system are mockups for collision detection and orientation only because of their insignificant impact on simulations. *Right:* Cut through the detector head with holder *(red)* and crystal *(orange)*.

with the according plug-in. In context of ongoing detector optimization, model updates are as simple as replacing the geometry file.

The media of the individual detector parts have been converted to V2M materials and an according resource was created. A simple taxonomy for segment-material mapping was created. Its main purpose is the identification of any detector's active volume. Parameters describing energy resolution and energy binning are stored in a tally resource (listing A.2). The association between tally and equipment is created by the scenario.

14.1.2 Stretcher

The stretcher is a customized massage table with individually adjustable segments for upper body, upper legs, and lower legs. Several configurations are predefined for different measurement setups. A comparison of stretched and sitting configurations was performed by Marzocchi et al. (2011) to determine an optimal setup.

The stretcher was modelled by Laubersheimer (2012) from scratch with SIMPLEGEO in three versions corresponding to the predefined setups used in measurements. However, the reclined version is used in most cases within the scope of this work.

14.1.3 Measurement chamber

The measurement chamber is shielded by 15 cm thick walls and a 25 cm thick floor of low-background steel (without impurities due to nuclear explosions). Additionally, the interior has a graded-Z shielding consisting of 5 mm lead (Z=82), 1.5 mm tin (Z=50), and 0.5 mm copper (Z=29).

A SIMPLEGEO model of the measurement chamber was adapted from an MCNPX version created by Hegenbart (2009).

14.2 Phantoms

Several phantom structures must be identified by VOXEL2MCNP's algorithms for specifying source locations, detector positions, and anthropometric parameters. This is done with a taxonomy common to all phantoms. Each segment in the geometry of an equipment (listing A.3) and each material in a materials collection is associated with at least one term defining the semantics of the item. This allows a segment-material mapping.

An according taxonomy (figure 14.4) consisting of 322 terms was created in context of the available phantoms and based on Medical Subject Headings (MESH) (U.S. National Library of Medicine 2012) for anatomic system classification and body region classification, and Terminologia Anatomica (Federative Committee on Anatomical Terminology 1998) and Terminologia Histologica (Federative International Committee on Anatomical Terminology 2007) for anatomic system classification and histologic classification respectively.

A materials resource (listing A.4) with 38 organic and anorganic media in the human body was defined common to all phantoms according to specifications from ICRU (1992a), ICRP (2002), and ICRP (2009). Additional resources were created for physical phantoms according to their specifications of tissue-equivalent media.

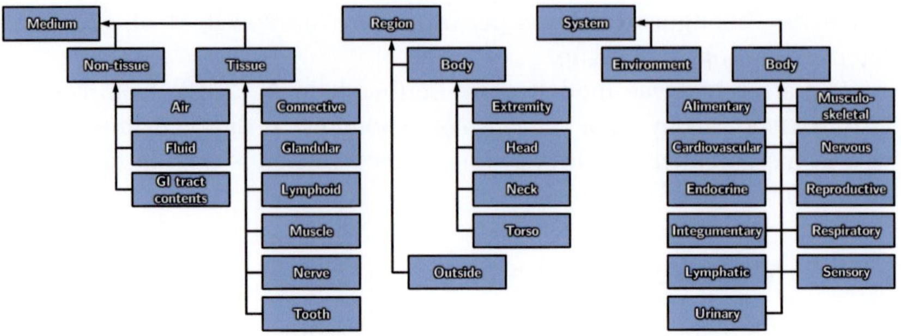

Figure 14.4: First three levels of the designed phantom taxonomy with medium, region and system classification. Each geometry segment and each material is associated with multiple, specific terms to define its semantics, which is used for identification and segment-material mapping.

Having defined taxonomy, equipment and materials, an initial segment-material mapping was generated with the method provided by VOXEL2MCNP (section 12.3), and then manually corrected by identifying structures based on their shape and relative location using the perspective view, and recording the information with the editor.

14.3 Scenarios

A scenario (listing A.1) combines multiple equipment resources and arranges them in space. These are up to four detectors, a phantom, the stretcher and the measurement chamber. A template scenario was created for the IVM body counter for each of the two main measurement setups. Instantiation of the template involves replacing the phantom, adding a source, and adjusting detector positions (figure 12.1).

14.3.1 Source emissions

26 energies covering the operational energy range of the detectors were selected and stored in a emissions resource (listing A.5). These values are used to sample typical calibration curves of the HPGe detectors with higher density at lower energies.

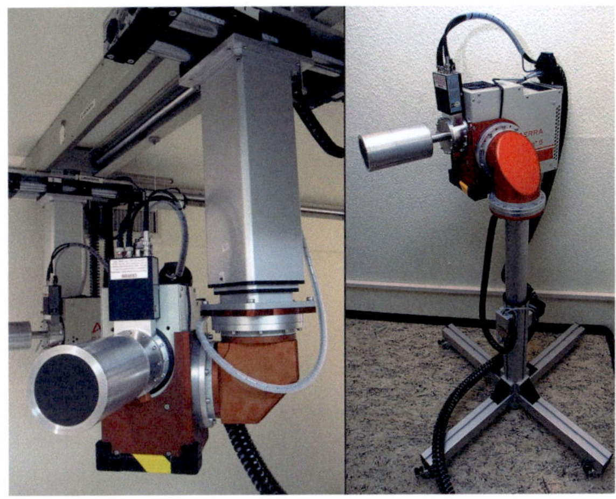

Figure 14.5: Custom detector mounting for ceiling and floor providing five degrees of freedom: translation in three directions, and yaw and pitch rotation.

Sources containing real radionuclides are modelled on demand via ENSDF files.

14.3.2 Detector positioning

Each detector features a custom mounting (figure 14.5) providing five degrees of freedom: translation in three directions, and pitch and yaw rotation. Detectors 1 and 2 are attached to racks with wheels that can be moved on the ground around the stretcher to access the person from the side. Detectors 3 and 4 are attached to a rail on the ceiling providing access from above.

A position recording system is available to record measurement setups for virtual reconstruction and subsequent simulation. It consists of a control rig that allows tracking and setting of detector positions and orientations. Additional dimensions are measured with a laser range finder relative to the walls or the floor. A previous version of the system that is in operation for the phoswich system is described by Hegenbart and Breustedt (2011).

The direct and inverse kinematic has been implemented in VOXEL2MCNP according to the mechanics of the mounting devices. This allows a 1:1 transfer of measurement setups between the physical and the virtual world.

Two main measurement setups were defined by Marzocchi (2011) for the body counter each using all four detectors (table 14.1): (1) 2×lungs, liver, and knee, and (2) 4×head. However, the actual positions are not fully specified and left to the interpretation of the technical staff performing the measurements. The default stretcher configuration for both setups has the upper body segment reclined to 45° and both leg segments slightly angled to improve personal comfort during long measurements.

To reduce modelling effort, improve reproducibility, and remove individual bias, an automated detector positioning strategy was implemented in the VOXEL2MCNP positioning module with the following scheme:

1. Assign a role to each detector predefined for each measurement setup.

2. Identify anatomical structures related to the role using the phantom taxonomy.

3. Determine orientation, dimensions and characteristic anatomical landmarks of the structures using principal component analysis.

4. Compile all information into parameters for the detector kinematic.

5. Adjust the distance to the body surface to a defined value (1 cm is the default value to account for involuntary motion during measurements).

6. Store the resulting transformation in association with the equipment object in the scenario.

Due to different segmentations of the various phantoms, some structures may not be available in all phantoms. Therefore, the implementation was designed to degrade with segmentation by using similar available structures and heuristically estimating certain locations from anatomical background knowledge. In some cases, it may also be possible to perform basic segmentation steps to properly identify certain segments.

Lungs-liver-knee

The detector roles are (1) liver, (2) left knee, (3) left lung, and (4) right lung (figures 14.6 and 14.7).

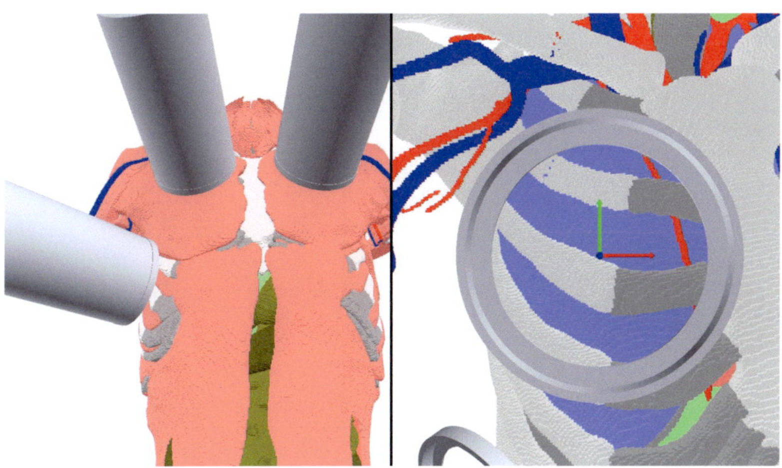

Figure 14.6: Detector positions for left lung, right lung and liver *(left)*, and detail of the right lung position at height of the third rib *(right)*. Some tissues have been removed for visualization only.

- *Liver:* At the height of the seventh rib on the right side of the body (between inferior end of the rib cage and infrasternal notch) with pitch of 45° and orientation in medial direction of the body.
- *Left knee:* Centred on the distal femur head at height of the distal end of the patella on the left leg with pitch of 45° and orientation in medial direction to the body.
- *Left/right lung:* At the height of the third rib (below the sternal angle) on the left/right side of the thorax with the entrance window close to but not covering the sternum. The detectors are perpendicular to the chest surface defined by the manubrium and upper half of the body of sternum with a pitch in the range of 15° to 25° and yaw of ±25°.

It was observed that liver position has a large variation in superior-inferior direction depending on the size of the lungs and is not consistently indicated by anatomic landmarks. The seventh rib approximates the mean liver centre in the available phantoms. An option for personalisation would be to apply abdominal percussion to identify the inferior end of the lungs. This, however, requires the expertise of medical staff members.

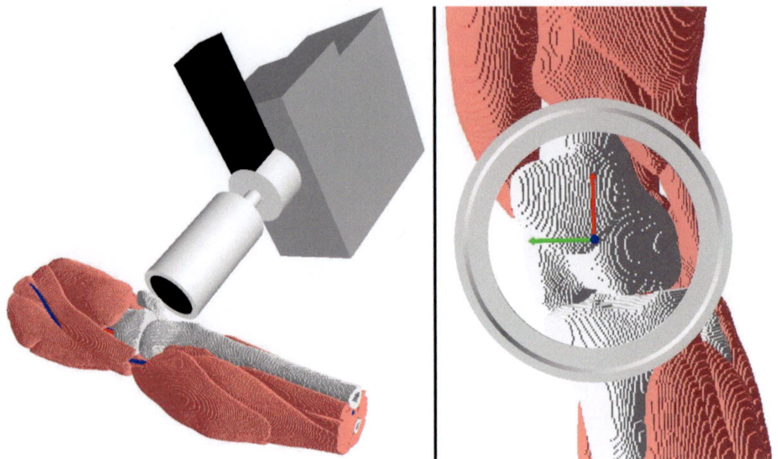

Figure 14.7: Detector positions for the left knee. The detector targets the distal end of the femur at the patella. Some tissues have been removed for visualization only.

Head

The detector roles are (1) back right, (2) back left, (3) front left, and (4) front right (figure 14.8). They are arranged at the sides of the head to avoid blocking the view of the person in case of anxiety. The front detectors are centred on the cranial suture (between the frontal and parietal bones) with the front casing beginning at the height of the eyebrows with pitch of $15°$ and yaw of $\pm70°$. The back detectors are centred on the left and right parietal bones with pitch of $-25°$.

14.3.3 Source locations

Similar to the association of detectors and tallies, phantoms are associated with sources (table 14.1). Four sources were defined corresponding to the two measurement setups by associating terms of the taxonomy.

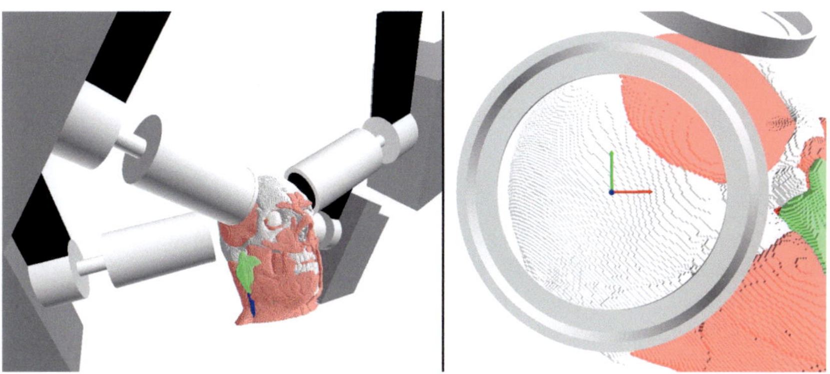

Figure 14.8: Detector positions for the head with detectors arranged in a cross *(left)*, and detail of the back right detector at the parietal bone *(right)*. Some tissues have been removed for visualization only.

Setup	Role	Source	Tally
Lungs-liver-knee	Liver	Liver	Detector 1
	Left knee	Skeleton	Detector 2
	Left lung	Left lung	Detector 3
	Right lung	Right lung	Detector 4
Head	Back right	Skeleton	Detector 1
	Back left	Skeleton	Detector 2
	Front left	Skeleton	Detector 3
	Front right	Skeleton	Detector 4

Table 14.1: Overview of roles and associated sources and tallies for the two considered measurement setups. Sources are associated with a set of phantom segments. Tallies are associated with the segment corresponding to the active volume of the detector crystal.

14.4 Simulation

All simulations have been performed according to the workflow depicted in figure 12.4. The MCNPX code (Pelowitz 2007) in version 2.7c was used as the radiation transport code for all simulations.

14.4.1 Preprocessing

In some cases, preprocessing of the created scenarios is necessary to improve computational efficiency or even allow computation in the first place in case of exceeded memory requirements. This may be done by replacing models with simplified versions or removing them if only insignificant impact on the simulation results is expected.

A phantom may be replaced with a simplified version consisting only of a region around the source. This may also remove a part of the source, which makes additional normalization in post-processing necessary. Obviously, crosstalk efficiencies cannot be evaluated in this case. The stretcher, measurement chamber, and mockup parts of the detectors may be removed completely.

After preprocessing, conversion to MCNPX syntax was done with the according export plug-in.

14.4.2 External simulation

All simulations were performed on a high performance cluster at KIT Steinbuch Centre for Computing (SCC) or at local machines at Institute for Nuclear Waste Disposal (INE). Simulation runs were generally terminated when the tallies achieved a relative error of $< 1\,\%$. Tally fluctuation charts were checked to ensure that all tallies comply with the ten statistical tests (Shultis and Faw 2006) imposed by MCNPX.

14.4.3 Post-processing

The output files were imported with the according plug-in. Post-processing consisted of peak analysis for the expected photo peaks using the step background model (equation 3.5b), correction for cropped source volumes, and aggregation of left and right lung detectors and sources as well as the head detectors to remove redundancies and provide more robust data.

The resulting parameters were serialized to V2M SCHEMA and exported as tabular data for further evaluation.

Bone volume correction

Given the source region B for the skeleton and a conservatively cropped version B^* with counting efficiency $\eta_{D_i \leftarrow B^*}$ with respect to detector D_i. The count rate of the detector cps_{D_i} is assumed to be equivalent for both sources.

$$cps_{D_i} = \eta_{D_i \leftarrow B}\, A_B = \eta_{D_i \leftarrow B^*}\, A_{B^*} \tag{14.1}$$

The total source activity A_B is assumed to be homogeneously distributed over its volume V_B.

$$\frac{A_B}{V_B} = \frac{A_{B^*}}{V_{B^*}} \tag{14.2}$$

It follows that the full counting efficiency $\eta_{D_i \leftarrow B}$ is equivalent to the volume-weighted partial counting efficiency $\eta_{D_i \leftarrow B^*}$.

$$\eta_{D_i \leftarrow B} = \frac{V_{B^*}}{V_B}\, \eta_{D_i \leftarrow B^*} \tag{14.3}$$

Combination of lung sources and detectors

Given a set of linear equations relating detectors D_1, D_2, D_3, and D_4 arranged at liver, left knee, left lung, and right lung and their respective sources LV, B, LL and RL.

$$\begin{pmatrix} cps_{D_1} \\ cps_{D_2} \\ cps_{D_3} \\ cps_{D_4} \end{pmatrix} = \begin{pmatrix} \eta_{D_1 \leftarrow LV} & \eta_{D_1 \leftarrow B} & \eta_{D_1 \leftarrow LL} & \eta_{D_1 \leftarrow RL} \\ \eta_{D_2 \leftarrow LV} & \eta_{D_2 \leftarrow B} & \eta_{D_2 \leftarrow LL} & \eta_{D_2 \leftarrow RL} \\ \eta_{D_3 \leftarrow LV} & \eta_{D_3 \leftarrow B} & \eta_{D_3 \leftarrow LL} & \eta_{D_3 \leftarrow RL} \\ \eta_{D_4 \leftarrow LV} & \eta_{D_4 \leftarrow B} & \eta_{D_4 \leftarrow LL} & \eta_{D_4 \leftarrow RL} \end{pmatrix} \begin{pmatrix} A_{LV} \\ A_B \\ A_{LL} \\ A_{RL} \end{pmatrix} \tag{14.4}$$

The total lung source activity $A_{LL} + A_{RL}$ is assumed to be homogeneously distributed over the volumes V_{LL} and V_{RL} of both lungs.

$$\frac{A_{LL} + A_{RL}}{V_{LL} + V_{RL}} = \frac{A_{LL}}{V_{LL}} = \frac{A_{RL}}{V_{RL}} \tag{14.5}$$

Summing the count rates of left lung and right lung results in a virtual detector $D_{3,4}$ and a virtual source (LL, RL) with counting efficiency equivalent to the volume-weighted sum of their respective (volume-corrected) counting efficiencies with weights $w_{LL} = V_{LL}/(V_{LL} + V_{RL})$ and $w_{RL} = V_{RL}/(V_{LL} + V_{RL})$.

$$\eta_{D_{3,4}\leftarrow LV} = \eta_{D_3\leftarrow LV} + \eta_{D_4\leftarrow LV} \tag{14.6a}$$

$$\eta_{D_{3,4}\leftarrow B} = \eta_{D_3\leftarrow B} + \eta_{D_4\leftarrow B} \tag{14.6b}$$

$$\eta_{D_1\leftarrow LL,RL} = w_{LL}\,\eta_{D_1\leftarrow LL} + w_{RL}\,\eta_{D_1\leftarrow RL} \tag{14.6c}$$

$$\eta_{D_2\leftarrow LL,RL} = w_{LL}\,\eta_{D_2\leftarrow LL} + w_{RL}\,\eta_{D_2\leftarrow RL} \tag{14.6d}$$

$$\eta_{D_{3,4}\leftarrow LL,RL} = w_{LL}\left(\eta_{D_3\leftarrow LL} + \eta_{D_4\leftarrow LL}\right)$$
$$+ w_{RL}\left(\eta_{D_3\leftarrow RL} + \eta_{D_4\leftarrow RL}\right) \tag{14.6e}$$

Combination of head detectors

Given a set of linear equations relating detectors D_1, D_2, D_3, and D_4 arranged at the head and the skeleton source B.

$$\begin{pmatrix} cps_{D_1} \\ cps_{D_2} \\ cps_{D_3} \\ cps_{D_4} \end{pmatrix} = \begin{pmatrix} \eta_{D_1\leftarrow B} \\ \eta_{D_2\leftarrow B} \\ \eta_{D_3\leftarrow B} \\ \eta_{D_4\leftarrow B} \end{pmatrix} A_B \tag{14.7}$$

Summing the count rates of all detectors results in a virtual detector $D_{1,2,3,4}$ with counting efficiency equivalent to the sum of their respective (volume-corrected) counting efficiencies.

$$\eta_{D_{1,2,3,4}\leftarrow B} = \eta_{D_1\leftarrow B} + \eta_{D_2\leftarrow B} + \eta_{D_3\leftarrow B} + \eta_{D_4\leftarrow B} \tag{14.8}$$

15 Measurement reconstruction

Hegenbart (2009) already performed measurements and simulations of the LLNL phantom in liver and lung counting setups and showed the validity of the approach of computational body counter calibration for the IVM. To check for any issues with the redesigned VOXEL2MCNP, it was applied to the reconstruction of calibration measurements with the JAERI phantom (Shirotani 1988).

15.1 Phantom modelling

The JAERI phantom was created by Shirotani (1988) at the Japan Atomic Energy Research Institute for the purpose of application to body counter calibration for lung and liver counting setups with regard to transuranic radionuclides. It is based on the torso of a Japanese adult male with average body mass (63.5 kg), height (168 cm), and chest circumference (90.5 cm). The phantom has an artificial ribcage with lungs, heart, liver, kidneys, and chest plate. It is provided with three sets of chest overlays with different muscle-adipose ratios (10:90, 20:80 and 30:70). Each set consists of two overlays with varying thickness (0.8 cm and 1.5 cm). The base phantom has an average chest wall thickness of 1.5 cm.

Voxel models of the physical phantom were constructed by Laubersheimer (2011) (figure 15.1) using computed tomography and image segmentation of the phantom with inactive lungs and both types of chest overlays. Segmentation was performed with the image segmentation software tools OSIRIX (Rosset, Spadola and Ratib 2004) and MIMICS (Materialise NV 2013). The segmented slices were stored as image series and converted with IMAGEJ (Ferreira and Rasband 2012) to a binary data format, which can be read by VOXEL2MCNP.

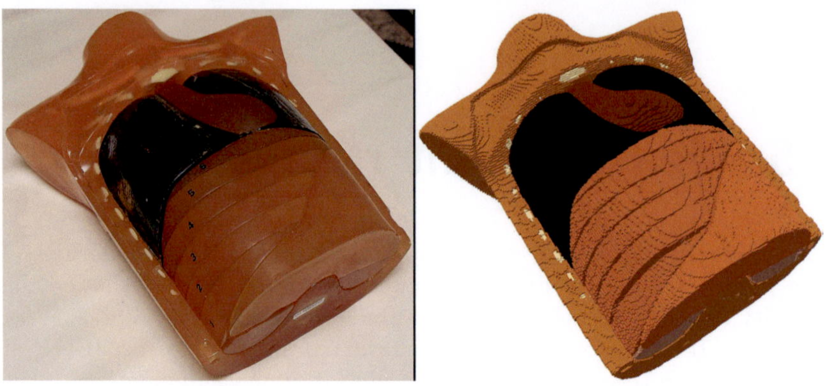

Figure 15.1: Photograph and geometric model of the JAERI phantom with removed chest cover (Pölz et al. 2013). The blank lungs can be replaced with active versions. The liver can be loaded by inserting special sheets between the individual slices (not available for this work).

15.2 Source modelling

The JAERI phantom was provided with five active lung sets: ^{241}Am (IAEA-AM2), $2\times^{238}$Pu (IAEA-PU5 and IAEA-PU9), ^{238}U enriched with 3% ^{235}U (IAEA-5U-5), and ^{232}Th (IAEA-TH7). According photon emission spectra were taken for all radionuclides and their relevant progeny in form of ENSDF from Laboratoire National Henri Becquerel (2013). In addition, X-ray emissions were taken from NUDAT (National Nuclear Data Center 2013b) and TORI (Firestone and Ekström 2004) databases. ENSDF files were imported with a file plug-in to VOXEL2MCNP and then reviewed and extended. Emissions with very low contribution were discarded upon MCNPX code generation using a threshold filter to improve computational efficiency.

15.3 Detector positioning

Three HPGe detectors were used in the measurements and positioned above left lung, right lung, and liver (figure 15.2). For enhancing reproducibility of the positions, the phantom was adjusted on the stretcher with

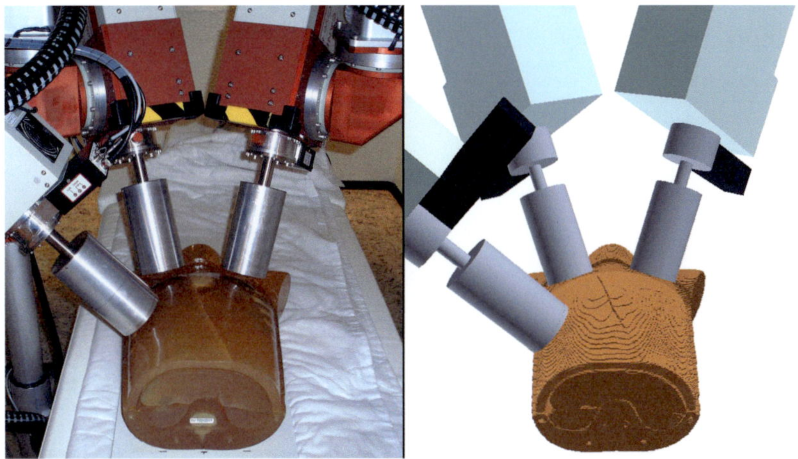

Figure 15.2: Photograph and geometric model of the calibration measurements with the JAERI phantom at IVM (Pölz et al. 2013). Two detectors are targeting the active lungs and one the inactive liver.

the help of a laser rangefinder and the positioning of the detectors was done with the electronic position recording system.

15.4 Results and discussion

In total, 21 measurements were taken (two examples are given in figure 15.3), including two background measurements with blank lungs. The software GENIE 2000 (Canberra 2006) was used for evaluation of the measured spectra, i.e. peak localization and peak area estimation. The results of the determined activities are presented in table 15.1.

Some measurements are exactly reproduced by their according simulations and others have large deviations. These deviations are consistent among lung sets in the estimated activity and may be due to three main factors:

- Missing details in the phantom model, such as air inclusions in the lung material of the active lung set or any damages that happened after imaging,

Lung set	Reference year	Reference activity in Bq	Radio-nuclide	Relative activity in %
IAEA-AM2	1993	420	^{241}Am	97.00
IAEA-PU9	1995	5000	^{238}Pu	87.44
IAEA-PU5	1996	42 300	^{238}Pu	88.13
IAEA-5U-5	1987	1676	^{238}U	100.00
			^{235}U	100.00
IAEA-TH7	1963	125.8	^{232}Th	100.00

Table 15.1: Overview of activities determined for several JAERI lung sets as ratio of simulation and measurement (Laubersheimer 2012). Percentages of decay products of uranium and thorium were also calculated.

- Imprecise detector positioning due to tolerances in mechanics and measurements (± 0.5 cm in translations and $\pm 1°$ in rotations), and

- Uncertainty in certified activities and possible inhomogeneous distribution inside the lungs.

Attempts to virtually fix several parts of the phantom model had no significant impact on counting efficiency. The uncertainty due to imprecise detector positioning was estimated to about 5 % at the peak efficiency using parametric simulations. These results are similar to values reported by Hegenbart (2009) for phoswich detectors. Changes to the lung activities, which scale counting efficiencies of multiple peaks, result in a much better shape of the calibration curve over all lung sets. This is an indication that there may be inhomogeneities due to technical difficulties in the production process of the phantom lungs. This was analysed by Hegenbart (2009) for the LLNL phantom. The author noticed a systematic increase in the lung density towards the boundaries, which could be due to a special coating to prevent abrasion.

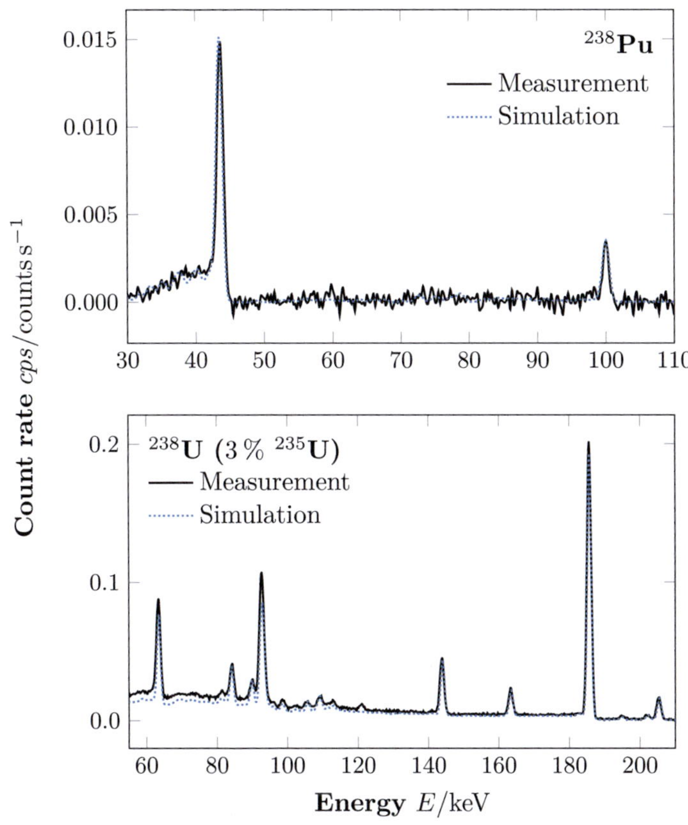

Figure 15.3: Comparison of peak shapes for measurement and simulation of the JAERI phantom with the ^{238}Pu and ^{238}U enriched with 3 % ^{235}U lung sets. The count rates are given for the combination of two detectors. The measurement time was about 14 h in both cases.

16 ICRP-89 phantoms

The ICRP reference man specification (ICRP 2002) provides values of body height, body mass, and organ masses for males and females of various age groups besides many values for physiological quantities for application to radiation dosimetry. This specification has been implemented by four research groups in form of computational phantoms. However, the official reference computational phantoms are the ICRP phantoms (ICRP 2009).

As a part of this work, all phantoms were used as calibration phantoms for the IVM body counter to quantify the order of magnitude of uncertainty that is produced by anatomic features not described by the specification. These features are primarily body and organ shapes that are defined by the base data used for phantom construction.

16.1 Phantom modelling

The following phantom series each consisting of a pair of adult male and female models were acquired and imported into VOXEL2MCNP:

ICRP (ICRP 2009): The phantoms have a lattice structure based on computed tomography data of two individuals with near-reference body mass and height. The segmented structures were modified using voxel modification methods. The phantoms are designated ICRP-AM (male) and ICRP-AF (female).

RPI (Zhang et al. 2009): The models are based on polygonal mesh surfaces from a database of anatomical structures. The authors combined the models using a deformation algorithm to resolve volume overlap. The phantoms are designated RPI-AM (male) and RPI-AF (female).

UFPE (Cassola et al. 2010): The models are based on polygonal and NURBS meshes from a database of anatomical structures and a parametric body surface model. Geometric modelling tools were used to adjust and combine the different parts. The phantoms are designated MASH (male) and FASH (female).

UFH (Lee et al. 2007; Lee et al. 2010): The models are based on polygon and NURBS meshes approximating segmented computed tomography data. Additional reference anthropometric parameters were derived from a nutrition and health survey and used for the creation. The phantoms are designated UFHADM (male) and UFHADF (female).

16.2 Anthropometry and organ masses

The adult male and female versions conform in general to the specified body heights (176 cm and 163 cm) and body masses (73 kg and 60 kg). However, there are issues with organ masses of several structures (table 16.1). While many are perfectly represented (e.g. heart, brain, kidneys, and pancreas), there are others with significant deviations (e.g. adipose, muscle, and lung tissue) that cannot be explained by different modelling techniques and representation methods including volume changes due to low voxel resolution.

The authors of the ICRP series state that the imaging modality — computed tomography in supine position — caused compression of the lungs for both persons. The lung density was therefore increased from 0.25 to $0.38 \, \mathrm{g/cm^3}$ to compensate the low volume with regard to radiation transport. Similar modifications to $0.36 \, \mathrm{g/cm^3}$ for the UFH series and to $0.27 \, \mathrm{g/cm^3}$ for UFPE were applied for consistency among all phantoms. The RPI series did not need any modifications. These differences are systematic, since they are present in both male and female phantoms of the series. It was ensured that the computed volumes include the full lungs with segmented blood vessels as specified by ICRP. Inclusion or exclusion of segmented blood vessels and bronchi cannot explain the differences.

16.3 Detector positioning

Both setups, lungs-liver-knee and head, were used for all eight phantoms. Although using the automated detector positioning strategy, differences in the detector positions are obvious due to very different body shapes. A comparison (figure 16.1) of the phantoms shows large differences in the

Resolution	ICRP	RPI	UFPE	UFH
x/cm	0.21	0.25	0.12	0.10
y/cm	0.21	0.25	0.12	0.10
z/cm	0.80	0.25	0.12	0.10
Structure	**ICRP**	**RPI**	**UFPE**	**UFH**
Adipose tissue	+0.12	+0.13	−0.20	−0.28
Muscle tissue	0.00	0.00	0.00	+0.43
Lungs	−0.40	0.00	−0.08	−0.31
Liver	+0.01	+0.01	+0.01	−0.01
Cortical bone	0.00	0.00	−0.02	−0.01
Skin	+0.13	+1.42	0.00	−0.48

Table 16.1: Comparison of voxel resolution and organ volumes of male ICRP-89 phantoms. Given are relative deviations from the reference volumes derived from specified masses (ICRP 2002) and densities (ICRU 1992a).

local anatomy of the chest regarding ribs and cartilages. Also, the liver varies in shape and position.

16.4 Results and discussion

The resulting calibration curves show deviations relative to the ICRP phantoms for all setups (figure 16.2). The deviations are generally increasing with decreasing photon energy. The values of the ICRP phantoms are mostly higher than those of other phantoms. This could be explained due to the lean body structure of the ICRP phantoms. The high attenuation at the liver of RPI-AF and RPI-AM could be explained by a large portion of the liver extending to the left side of the body. The deviations are generally not correlated to a change in volumes. There is also no simple relationship when comparing changes in counting efficiencies of male and female phantoms. The changes may primarily be caused by local deviations in the tissues shielding the source organs.

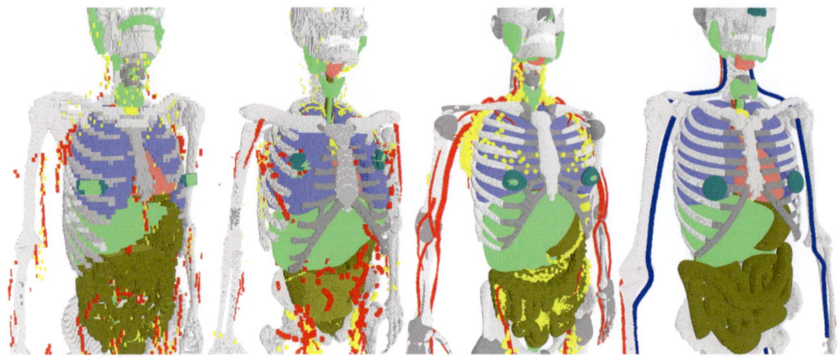

Figure 16.1: Comparison of anatomy for male ICRP-89 phantoms. *From left to right:* ICRP-AM, RPI-AM, MASH, and UFHADM. Visible are, bones and cartilages *(grey)*, liver and glandular tissue *(green)*, gastrointestinal tract *(brown)*, lungs *(light blue)*, muscle tissue and heart *(pink)*, blood vessels *(red/blue)*, and lymphatic tissue *(yellow)*. Skin, and major muscle and adipose tissues have been omitted for visualization.

It is obvious that there are large differences in phantom development starting by the used data over modelling bias to the representation method. These differences have an effect on three major factors:

- Anatomy and organ shapes

- Detector positions due to different body shapes

- Voxel resolution and segmentation

Differences due to changes in voxel resolution were quantified by Hegenbart (2009) for lung counting with phoswich detectors of the LLNL phantom to about 0.9 % per 1 mm. The influence of body shape is expected to be relatively large because of the high spatial sensitivity of the HPGe detectors. Deviations increase with decreasing energy, which makes body counter calibration very sensitive to anatomical changes for incorporation of low-energy photon emitters.

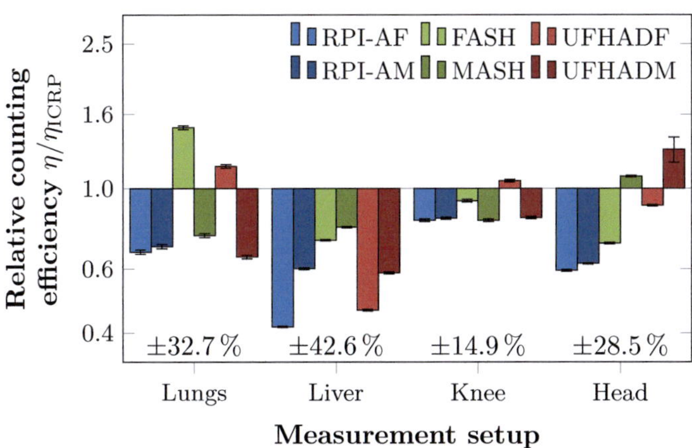

Figure 16.2: Comparison of counting efficiencies for all detector setups. The values are given for the main photo peak of ^{241}Am (59.5 keV) as a representative for low-energy photon emitters. The values are normalized to their gender-specific references from ICRP. The percentages at the bottom give the relative standard deviation of the values.

17 XCAT phantom series

The application of the STEP framework to a large phantom series is the main idea of this work. The XCAT series (Segars and Sturgeon 2010) was selected for this part, because it is the only available large series of person-specific phantoms with high anatomical detail.

17.1 Calibration

The XCAT series (Segars and Sturgeon 2010) is a set of 30 adult, person-specific, whole body phantoms derived from computed tomography of medical patients. They have a high level of anatomical detail and comprise 2724 individual structures (figure 17.1). The series is available as a set of NURBS files with an associated tool, called DXCAT2, which has several functionalities:

- Voxelization to a lattice of densities for medical tomographic imaging specified by in-plane resolution, plane offset, and gantry dimensions. There is also an option to produce segment identifiers instead.

- Placement of a spherical lesion for radiation treatment planning.

- Setting of activities in several major structures for emulating radioactive tracers nuclear medicine.

- Change of diameters and volumes of several major structures to adapt anatomical features.

- Adjustment of position in the breathing and heart motion cycles based on average interpolated data of individuals.

Voxelization was done with DXCAT2 for all phantoms to voxels with 1 mm edge length and associated segment identifiers. This seemed to provide a sufficient level of anatomical detail. For variance reduction, each phantom was cropped into three regions containing the head (figure 17.2),

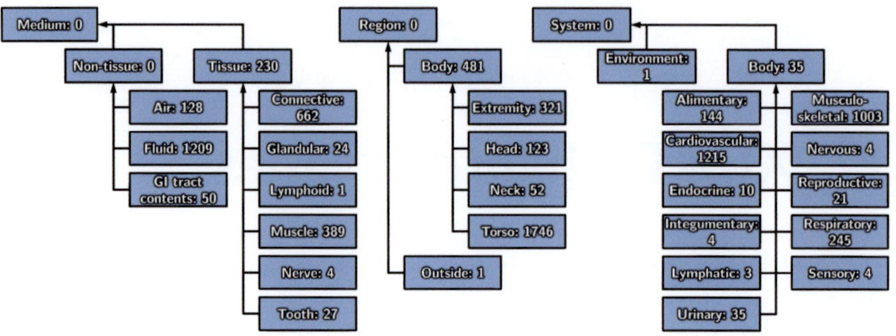

Figure 17.1: First three levels of the designed phantom taxonomy with medium, region and system classification, and number of registered XCAT segments.

torso (figure 17.3), and left knee (figure 17.4) based on the design of corresponding physical phantoms. This was done by defining cutting planes at certain relative positions on major bones. Also, the stretcher and mockup parts of the detectors needed to be removed. Hegenbart (2009) already showed that such details may have a large impact on point source measurements (a 26.2 % decrease of counting efficiency for ^{241}Am) due to missing scattering effects, but has virtually no effect for body counting.

Both setups, lungs-liver-knee and head, were used for all phantoms. The phantom parts were arranged according to the angled configuration of the stretcher. No organ shift was added to compensate the posture change from supine position. Simulations were performed for 30 phantoms, with five types of source locations, with 26 photon emission energies, and eight detector roles defined by the measurement setups. This leads to a total of 3900 simulations and 6240 tallies. 780 samples of counting efficiency remain for each source and primary tally after aggregation.

The resulting calibration curves are in similar ranges with the LLNL (Griffith et al. 1987) and JAERI (Shirotani 1988) phantoms. Deviations from the mean of all XCAT phantoms are decreasing with increasing photon energy. For an energy range from 25 keV to 2 MeV, relative standard deviation in counting efficiencies decreases from 61.1 % to 21.1 % for lungs, from 48.9 % to 21.1 % for liver, from 32.4 % to 15.1 % for knee, and from 15.8 % to 11.6 % for head. An example is given in figure 17.5.

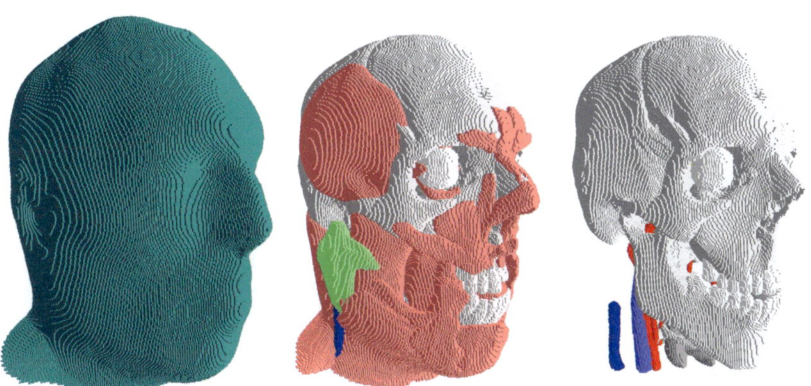

Figure 17.2: Structure of the head region of an XCAT phantom with adipose tissue (*left*), muscle tissue (*centre*), and skeleton and internal organs (*right*).

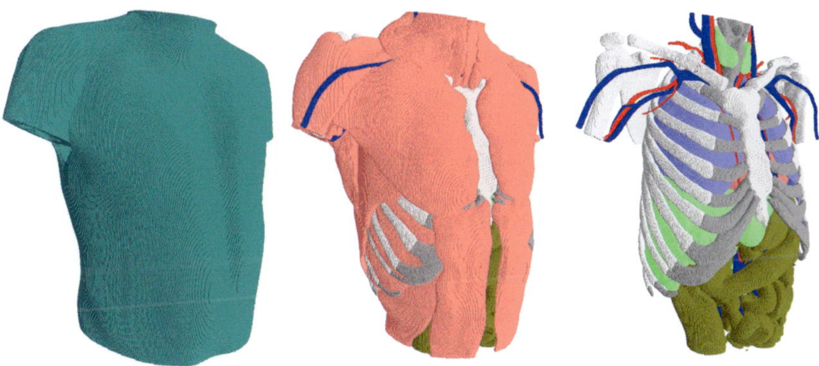

Figure 17.3: Structure of the torso region of an XCAT phantom with adipose tissue (*left*), muscle tissue (*centre*), and skeleton and internal organs (*right*).

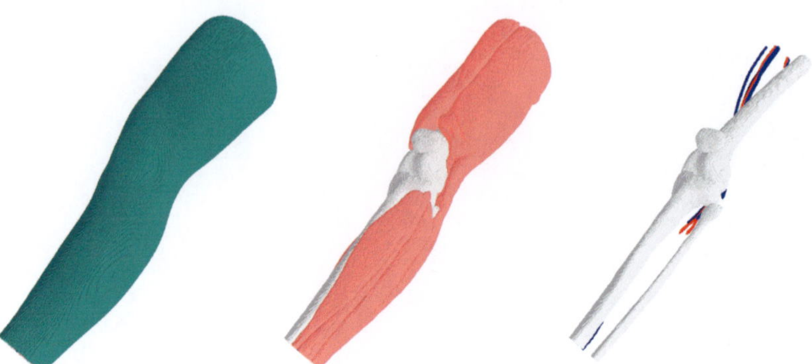

Figure 17.4: Structure of the knee region of an XCAT phantom with adipose tissue (*left*), muscle tissue (*centre*), and skeleton and internal organs (*right*).

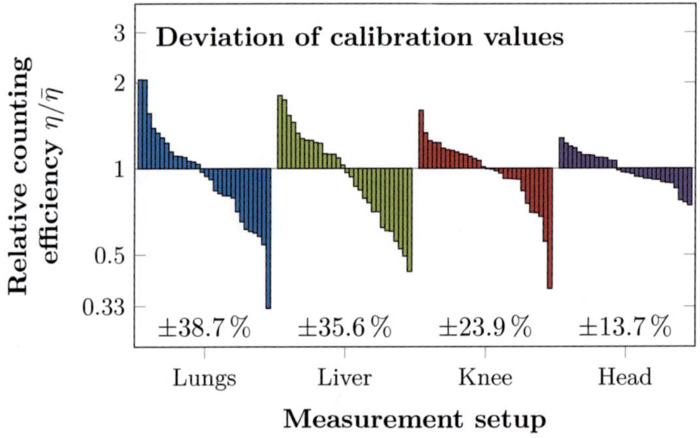

Figure 17.5: Relative standard deviation of XCAT calibration values for different measurement setups. The values are given for the main photo peak of ^{241}Am (59.5 keV) as a representative for low-energy photon emitters. The percentages at the bottom give the relative standard deviation of the values.

17.2 Anthropometry

Performing anthropometric measurements according to a specification is necessary to produce reproducible and comparable values for people as well as computational phantoms. The specification used in this work is EN 13402-1 (European Committee for Standardization 2001), which is the European standard for clothes sizes.

It was taken care that there are no systematic differences in the procedures for physical and computational measurements. The only differences are due to posture-related deformations of the person and aliasing effects due to the geometric representation.

In the following, a list of implemented measures with a brief description is given. Of course, not all implemented measures are useful for every detector setup. But, they can still be used as control parameters to check for possible deviations with respect to general expectations. For instance, a strong correlation between body height and inner leg length is expected, if it is not visible in the data, there may be systematic errors in the computation.

The considered anthropometric parameters are grouped into three categories: masses and derivatives (table B.1), lengths, breadths and distances (table B.2), and circumferences (table B.3). Details for physical and virtual measurements are indicated by the according keywords. Additionally, average photon transmission (section 12.4) was computed for each source organ and primary detector for each simulated photon energy.

All 18 anthropometric parameters were computed for each phantom leading to a total of 540 values. The parameters are comparable to values from the National Health and Nutrition Examination Study (NHANES) (Department of Health and Human Services and National Center for Health Statistics 1996), which gives an impression of the distribution of parameters (figure 17.6) among the U.S. population and clients of the IVM. Body masses of physical phantoms with overlays were extrapolated from the specified body masses of the base phantoms by linear scaling with the mass ratio gained through the overlay. Apparently, the definition of "chest circumference" for the IVM and in the specification of JAERI and LLNL phantom are closer to the definition of bust circumference than chest circumference with regard to this work (table B.3). Experimental measurements with a tape measure on the physical phantoms confirm this systematic difference in the definition.

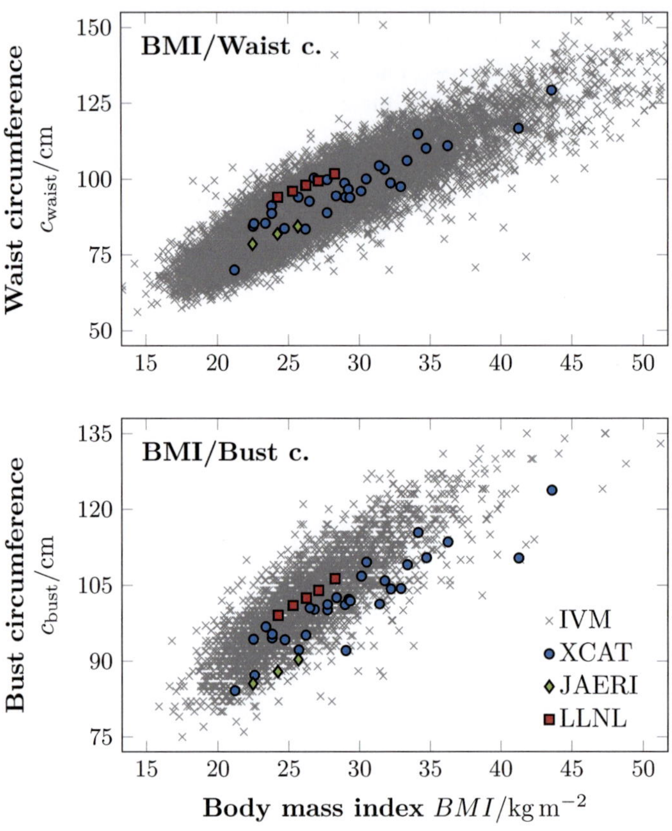

Figure 17.6: XCAT anthropometric parameter values in context of the LLNL and JAERI phantoms, and statistical data. The percentage of males in the XCAT data is 50.0 %. *Top:* 12 900 adults (aged 18 to 65) with 47.9 % males measured during NHANES (Department of Health and Human Services and National Center for Health Statistics 1996). *Bottom:* 3247 persons with 90.8 % males measured at IVM.

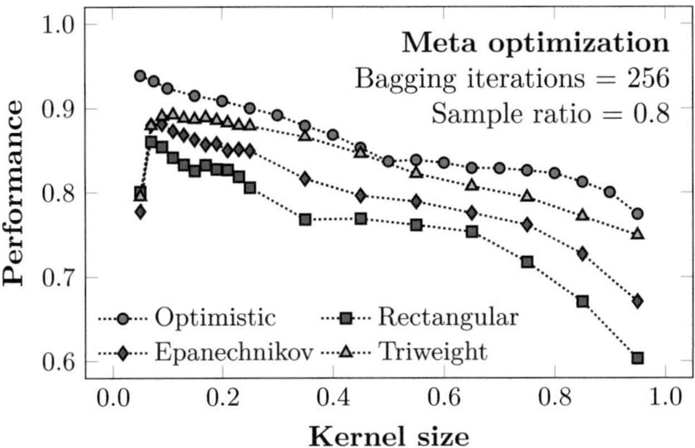

Figure 17.7: Optimization of free parameters for statistical analysis with bagging and local polynomial regression. The liver data set with a subset of energy and waist circumference was used as sample data. Lungs, knee, and head show similar results. For comparison, *optimistic* designates the performance without the use of bagging, which is obviously overfitting for low kernel sizes.

17.3 Statistical analysis

Statistical models were built according to the descriptions in chapter 13. The method for creating an estimator given a feature subset has several free parameters. Reasonable samples for these parameters were selected and optimized with regard to the performance on the data using grid optimization (figure 17.7):

Bagging iterations: The number of bagging iterations should be as large as possible to get robust performance estimates. The constraining factor is computer time. Values of $16, 32, \ldots, 256$ were tested. A value of 256 was estimated to provide a good tradeoff.

Sampling method: Bootstrapping was chosen as sampling method above cross sampling and random sampling.

Sample ratio: Sample ratio defines the size of the training data set relative to the total number of samples. The remaining samples are used as testing data. The ratio should be as large as possible without encouraging overfitting and while providing enough testing data. Performance is slightly increasing with sample ratio due to better fitting of the data and was set to 0.8 based on literature values.

Kernel method: Support vector regression was discarded because of problems with computational efficiency. Therefore, local polynomial regression of order two with Euclidean distance for neighbourhood weighting and 20 passes for reduction of variance in residuals was selected.

Smoothing kernel: Most smoothing kernels perform relatively well with Triweight being best and Rectangular being worst. Other considered kernels were Bisquare, Gaussian, Epanechnikov, Exponential, McLain, Triangular, and Tricube. They also exhibit similar behaviour with regard to other parameters.

Kernel size: Kernel size is relatively independent from other parameters and has a peak at 0.1. Values up to 0.25 are also comparable and provide less overfitting. Smaller values emphasize random properties of the data and higher values ignore samples at the boundaries of the sample space and details.

The error bars on performances and counting efficiency estimates in the diagrams of this chapter represent the standard deviation among the individual bagging models and therefore the variation of the data. This is a measure of the robustness of the estimator and is mostly dependent on the sample ratio. The standard deviation of the mean performance and counting efficiency is only dependent on the number of bagging iterations, which is very low for 256. The actual uncertainty of the estimate with respect to the estimator could not be derived with the available implementation of RAPIDMINER.

Preselection for feature subset selection was performed by computing the performances of all individual features and covariances for all feature pairs, and removing those with low relevance and high redundancy. Optimal feature subsets were computed with brute force search for direct contributions of lungs, liver, knee, and head for feature subsets up to size three with energy as a permanent feature. It was checked in all cases that residuals are uncorrelated and estimators produce characteristic calibration curves when plotted over the energy range.

17.4 Results and discussion

The results show that adding another feature to energy almost always increases the estimator's performance. Adding a third feature may increase or decrease performance though. Sporadic tests of larger subsets showed no further improvement. This is due to increasing sparseness with increasing subset size. There is also a slight positive correlation between the performance increases for a single feature being added to different subsets.

Optimal features (figure 17.8) are related to the body region of the source structure. These are circumferences of the chest for lungs, of the chest and abdomen for liver, of the extremities for knee, and of the head for head setups. Due to the high correlation of several anthropometric parameters there are many feature subsets that perform only slightly suboptimal. Whole-body features, such as body mass and derivatives, show comparable but lower performances. Features related to body height and other lengths or other body parts show low performances. No correlation ($R^2 < 0.05$ for XCAT for all photon energies) was observed between cup size (difference of bust circumference and underbust circumference) and counting efficiency for lungs, which was reported by Hegenbart et al. (2008) and Farah, Broggio and Franck (2010). However, larger detectors (phoswich detectors and a 2×2 HPGe array) and detector positions (frontal to the breasts in the first case) were chosen in these works, which could explain the deviations. Also, no correlation ($R^2 < 0.05$ for XCAT for all photon energies) was observed between inverse lung volume and counting efficiency for lungs as reported by Farah, Broggio and Franck (2011b). However, the authors state that this effect is due to the modelling process of the applied phantoms.

It is possible to derive basic rules from the estimators, when plotted for specific features. For example, counting efficiency decreases with increasing corresponding circumference for all measurement setups (figure 17.9). Moreover, body circumferences are negatively correlated ($R^2 \approx 0.4$ for XCAT at 59.5 keV) to average photon transmission for fixed photon energies, which is itself positively correlated ($R^2 \in [0.6 - 0.9]$ for XCAT at 59.5 keV) to counting efficiency. This means that an increase in body circumference is likely also an increase in wall thickness of tissues at the location, which shield the source organ. In conclusion, the low change of counting efficiency for head measurements could be explained by the low increase in tissue thickness at the head with increasing circumference.

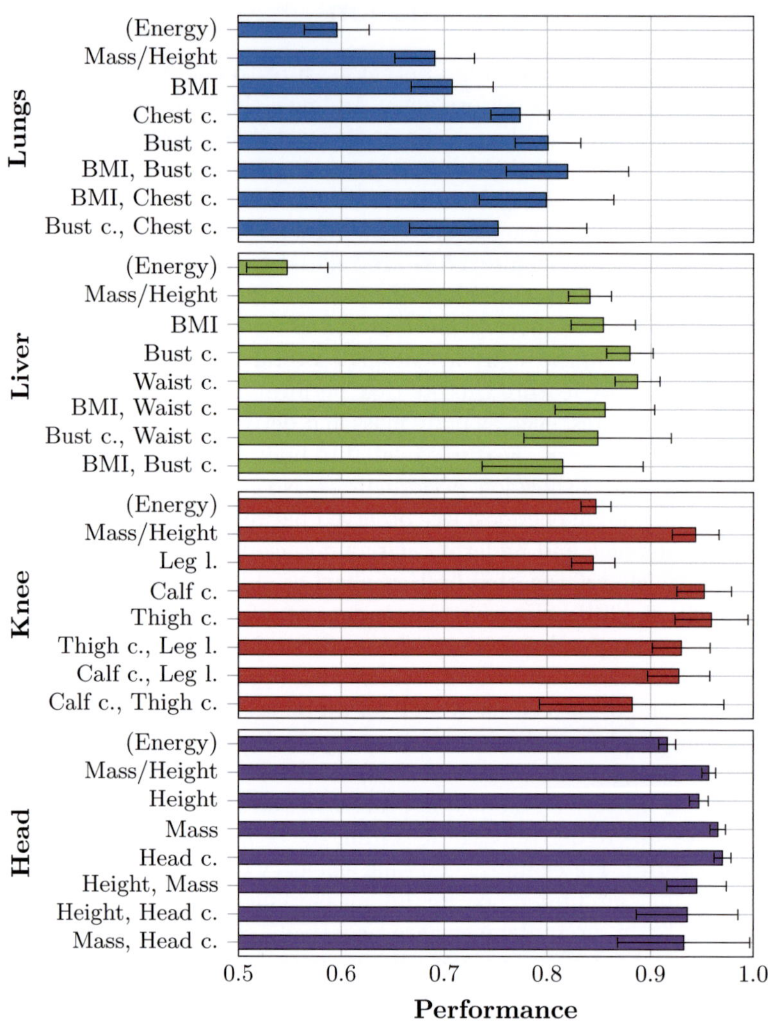

Figure 17.8: Performances of estimators trained with selected feature subsets. Energy is used as the base performance and present in each feature subset. The error bars show the standard deviation among the individual bagging models.

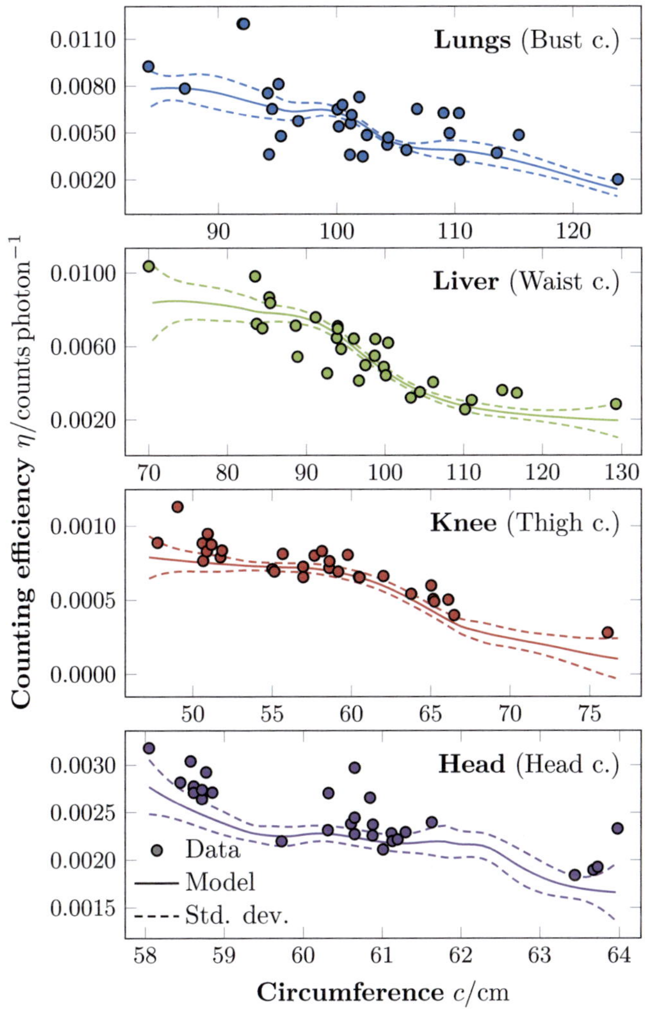

Figure 17.9: Models for subsets of energy and the corresponding sensitive circumference for all measurement setups. The curves are not based solely on the presented data points, but also on those for neighbouring energies. The values are given for the photo peak of ^{241}Am at $59.5\,\mathrm{keV}$ as a representative for low-energy photon emitters. The dashed lines show the standard deviation among the individual bagging models.

The legacy personalisation method (Mohr and Breustedt 2007) based on chest wall thickness currently applied at IVM uses a specific equation as a person model and the LLNL calibration phantom (Griffith et al. 1987) to construct a calibration model. For comparison to STEP, a calibration model of the LLNL phantom was created with VOXEL2MCNP and linear interpolation of average chest wall thickness with regard to the chest overlays of the phantom was performed for each photon energy. To evaluate the improvement of STEP compared to performing no personalisation and to the legacy personalisation method, three methods are proposed:

- Since STEP is an extension of the legacy method, it can emulate it using energy and the anthropometric parameter defined by ratio of body mass and height as feature subset (figure 17.8). The change in performance is a measure for improvement of the estimator. The score changes between the ratio of body mass and height and the optimal circumference relative to the base score are $+18.5\,\%$ for lungs, $+8.4\,\%$ for liver, $+1.8\,\%$ for knee, and $+1.5\,\%$ for head.

- A visual comparison of the goodness of fit of the legacy method and STEP (figure 17.10) shows an improvement for the XCAT and LLNL phantoms over the full energy range with regard to the squared correlation coefficient. This measure is biased towards the XCAT data for the STEP method and to the LLNL data for the IVM method when applied to the full data set.

- Another measure of the improvement is the change in relative standard deviation of the residuals of the estimators compared to the base estimator using only energy (figure 17.11). The changes for the optimal circumferences are $-9.0\,\%$ for lungs, $-16.5\,\%$ for liver, $-11.3\,\%$ for knee, and $-4.6\,\%$ for head. The corresponding values for the legacy method are $-4.0\,\%$ for lungs, and $-9.3\,\%$ for liver. Personalisation for head and knee cannot be performed with the associated torso phantom. This measure is biased towards the XCAT data for the STEP method, because it is evaluated on the data set that was also used for model training (figure 17.7).

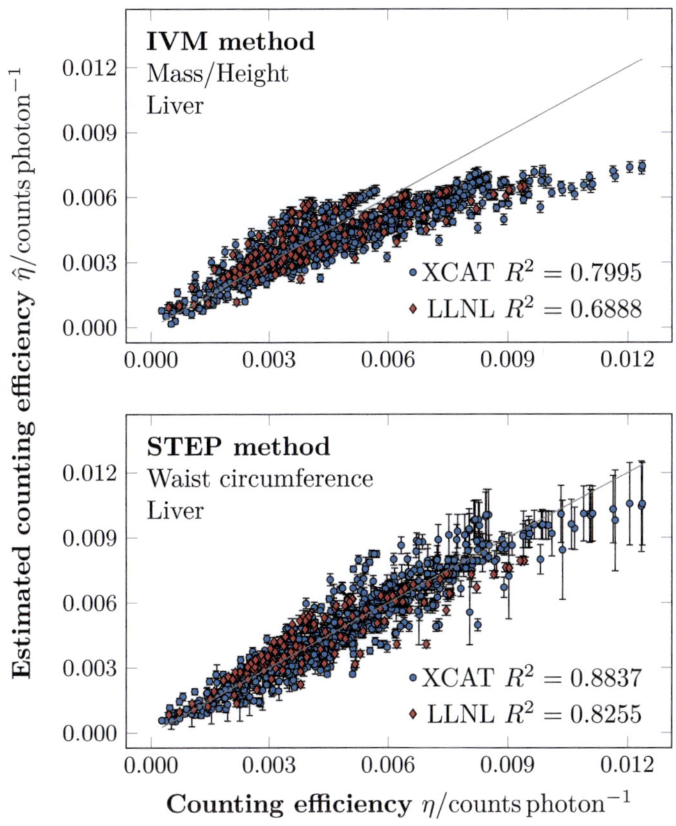

Figure 17.10: Comparison of estimated and computed counting efficiencies for the IVM and STEP method applied to XCAT and LLNL liver data. *Top:* IVM method with ratio of body mass and height. The standard deviations are based on the residuals of the linear regression of the LLNL data. *Bottom:* STEP method with photon energy and waist circumference. The error bars show the standard deviation among the individual bagging models.

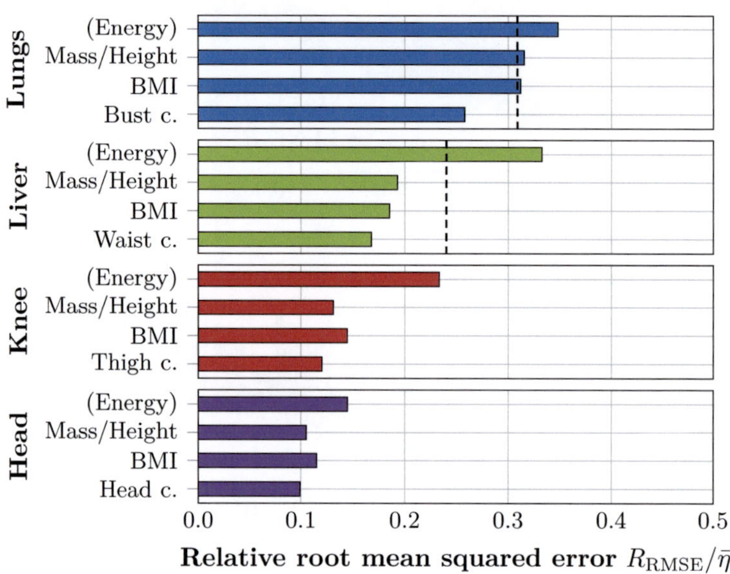

Figure 17.11: Relative root mean squared error of XCAT calibration values for different measurement setups and feature subsets. Results for the legacy method (*dashed line*) have been computed for the LLNL phantom. Energy is used as a base estimator.

18 Inhomogeneous source distributions

In general incorporation scenarios, radionuclides are not limited to one organ, but are a mixture of radionuclides over several organs, and even may be inhomogeneously distributed in those individual organs. The latter may be due to hot particles (Charles and Harrison 2007) that by themselves are a concentrated activity, or due to inhalation of radionuclides creating local deposition enhancement on bifurcations in the bronchioalveolar tree based on particle size or activity median aerodynamic diameter (Balashazy, Hofmann and Farkas 2002; Bergmann, Hofmann and Koblinger 1997; ICRP 1994).

Source distribution cannot be estimated with the usually available information and measurement systems. Therefore, the conventional assumption is a homogeneous distribution within an organ. However, source distribution has high sensitivity to efficiency calibration as shown by measurements (Pelled et al. 2006) and simulations (Kramer, Burns and Yiu 1997). It was shown that activities in lung counting can be underestimated by a factor of 20 and more for low-energy photon emitters. It was also shown that detector arrays reduce these errors by up to a factor of 4.

Measurements were performed as a part of this work to estimate the impact of inhomogeneous source distributions in lungs for the IVM body counter using a physical phantom with a set of perforated lungs and vials of ^{18}F.

18.1 Phantom modelling

The LLNL phantom was created by Griffith et al. (1987) at the Lawrence Livermore National Laboratories. Its structure and designated application are similar to the JAERI phantom. The main difference is that it is based on the torso of an average U.S. American adult males with respect to body mass, heights, and chest circumference.

Voxel models of the phantom were constructed by Hegenbart (2009). They were imported into VOXEL2MCNP as has been described in the case of the JAERI phantom (chapter 15).

18.2 Measurements and simulations

^{18}F-2-Fluor-2-deoxy-D-glucose (FDG) is a molecule containing radioactive ^{18}F. The radionuclide disintegrates with a half-life of 109.7 min by beta-plus decay (96.86 %) and by electron capture (3.14 %) to stable ^{18}O (Laboratoire National Henri Becquerel 2013). The main peak at 511 keV is produced by photons generated from electron-positron annihilation. The molecule is usually applied as a radioactive tracer in positron emission tomography.

The measurement setup consisted of the LLNL phantom with perforated lungs and two detectors targeting the lungs (figure 18.1). 20 vials filled with 10 µl FDG — equivalent to 51 270 Bq ^{18}F at reference time — were available. A variety of distributions were selected and measured. The setup was also measured with multiple chest overlays for the phantom using the same detector positions. In total, 17 measurements with ^{18}F were taken over the course of four hours with increasing measurement time from 1 to 10 min to compensate the fast decay. Additionally, a background measurement with the blank lung set was taken, and a reference measurement of the basic lung set with 84 200 Bq ^{241}Am was taken to check for any problems with the simulation parameters.

Five configurations were used for each lung (figure 18.2):

- Full loads with 14 vials left and 19 right

- Partial loads with five vials evenly distributed each in two variations

- Spot loads with one vial either in the front or in the back of the lung

The setup was also reconstructed and modelled with VOXEL2MCNP using homogeneous distribution of ^{18}F and ^{241}Am in lungs.

18.3 Results and discussion

Measured spectra were evaluated for the 511 keV peak of ^{18}F for both detectors (figure 18.3) and compared to the simulation results for the

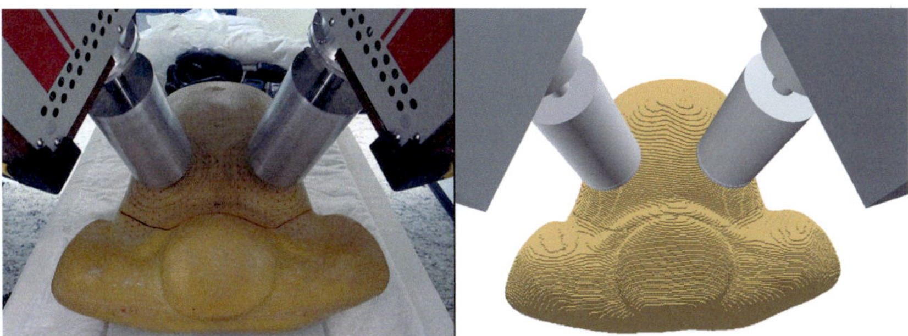

Figure 18.1: Physical and virtual measurement setup for the LLNL phantom with two detectors targeting the lungs.

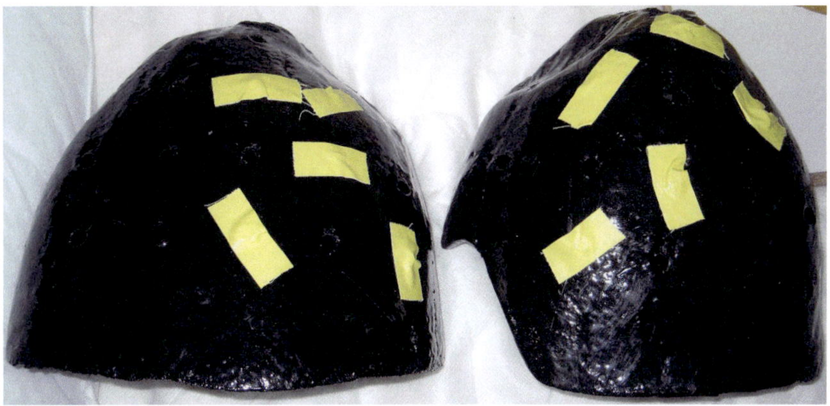

Figure 18.2: LLNL perforated lung set with left lung *(right)* and right lung *(left)* and several inserted vials. Different configurations were created by using a certain number of vials in a specific distribution. The front sides of the lungs are in the centre of the picture.

homogeneous distribution (figure 18.4). The full configuration is in 99.0($\pm$0.3)% agreement with the homogeneous distribution. Counting efficiencies for partial loads are 71.5($\pm$0.5)% and 61.2($\pm$0.7)% higher. This is probably due to the four spots in the back being empty in both configurations. The spot configurations are extremes. One vial in the back

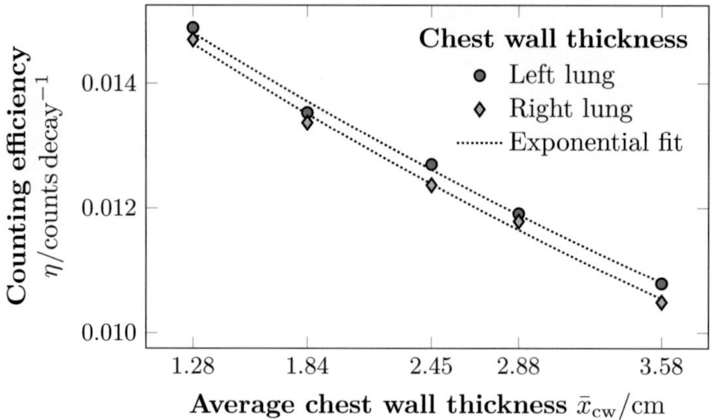

Figure 18.3: Dependency of measured counting efficiency and average chest wall thickness for partial configuration 1 with exponential fits ($R^2 > 0.99$).

produces only 35.7($\pm$0.3)% of the counts in the detectors and the front configuration produces 318.3($\pm$1.9)%. The difference between both front configurations may be due to the left spot being superior to the right spot and therefore closer to the detector (figure 18.2).

Pelled et al. (2006) reported factors of 7 for 185 keV and 10 for 92 keV. This is comparable to the factors determined in this work considering the different photon energies. Unfortunately, no additional measurements were possible due to the rapid decay of the radionuclide.

Simulation results of the reference measurement at the 59.5 keV peak of ^{241}Am agree to 95.8($\pm$1.0)% with regard to activity. This is an indication that the reconstruction parameters of the measurement setup are reasonable.

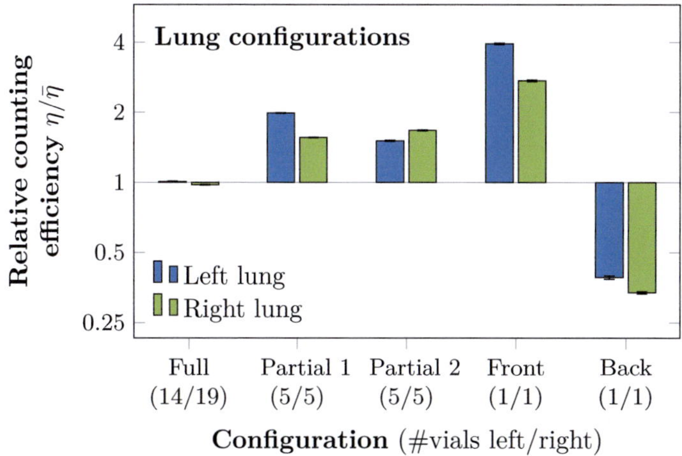

Figure 18.4: Comparison of measured counting efficiencies η for ten configurations relative to simulation results $\bar{\eta}$ for homogeneous source distribution of ^{18}F.

19 Respiratory motion

Complementary to the variation in anatomy between people, anatomy also varies intraindividually over short periods. An obvious variation is due to voluntary and involuntary motion.

Voluntary motion is expressed as a change in body posture when comparing body counting setups. This results in deformation of parts of the body. For partial-body setups with HPGe detectors, it is expected that there are minor variations in counting efficiencies due to posing. The largest contributions may be from crosstalk due to sources in muscle and bone. However, appropriate imaging data to perform comparisons is difficult to acquire since imaging is done primarily in supine position. Also, there are only few publications (Dimbylow and Findlay 2010; Nagaoka and Watanabe 2008) regarding posing in phantom development in radiation protection.

Involuntary motion stems from respiratory and cardiac motion. The expected variation due to these effects is also low, however, data is available for quantification in form of respiratory-correlated computed tomography data sets (Guckenberger et al. 2007).

19.1 Phantom modelling

Respiratory-correlated computed tomography data sets are usually used in image-guided radiotherapy to track small lesions in lungs and liver due to respiratory motion (Guckenberger et al. 2008). They are acquired with a multi-slice CT scanner combined with a pressure sensor fixed in the abdominal region of the patient. The sensor records pressure changes due to respiratory motion of the patient. The acquired images are sorted retrospectively by their associated position in the respiratory cycle and used for reconstruction of a series covering different phases of breathing (Guckenberger et al. 2007).

The available data sets have a spatial resolution of $0.7\,\mathrm{mm} \times 0.7\,\mathrm{mm} \times 2.0\,\mathrm{mm}$ and cover a cycle of tidal breathing in eight time steps. A selected

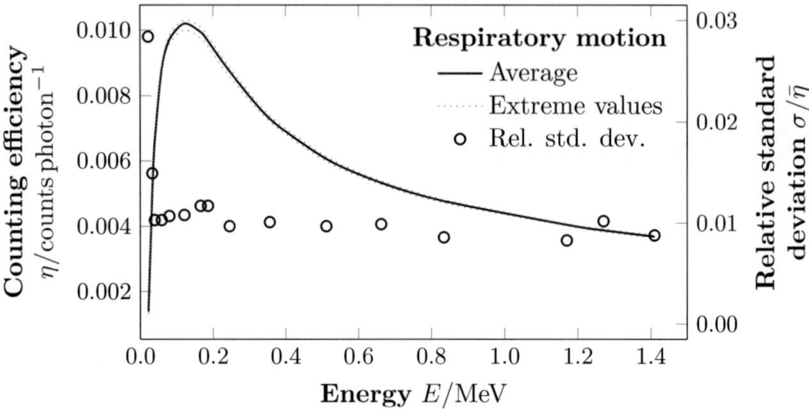

Figure 19.1: Computed counting efficiencies at full inspiration and expiration for the four-dimensional phantom and relative standard deviation from the average of all eight time steps. Expiration increases and inspiration decreases counting efficiency. The noise on the data is likely due to segmentation variance of the high-resolution data sets.

data set was segmented into six basic tissues (lungs, liver, bone, adipose, muscle, and soft tissue) and constructed into a four-dimensional phantom (Schneider 2011b). It was then applied to body counting with two fixed detectors over the lungs in all eight time steps (Schneider 2011a).

19.2 Results and discussion

The results (figure 19.1) show a standard deviation of counting efficiency over a full respiratory cycle of about 1 % for tidal breathing (Pölz et al. 2012). This is equivalent to a slight underestimation of counting efficiency for lung measurements when using a calibration phantom at full inspiration of tidal breathing. However, computed tomography data sets for phantom development are typically assessed at full inspiration of heavy breathing for better imaging of the thorax. Therefore, the observed effect on counting efficiency due to respiratory motion is expected to be larger considering those types of phantoms.

Part IV

Discussion

20 Summary and conclusion

The contribution of this work to body counter calibration and radiation protection in general is primarily the developed personalisation framework consisting of software implementation, the associated data model and a method for performing sensitivity analyses. The framework has been applied in several studies to analyse uncertainties with regard to the IVM body counter at KIT. However, it is not restricted to a particular facility or to body counting in general. It is designed to be extendable to other Monte Carlo codes and other applications related to dose assessment.

20.1 Development

The personalisation method currently applied at IVM (Mohr and Breustedt 2007) and similar body counting facilities (Lynch 2011; Pierrat et al. 2007) is based on estimation of chest wall thickness via the ratio of body mass and height. The major drawbacks are the missing variation in anatomy due to the use of a single physical calibration phantom, missing customization of the chest wall thickness estimation to specific body counting facilities, restriction to lung counting setups, missing estimation of uncertainties, and an overall low complexity of the model leading to coarse estimates (and incorrect extrapolation).

These drawbacks have been considered and improved with the design of the personalisation framework STEP (chapter 10). It is basically a strong generalization of the currently applied interpolation-based method. It considers a variety of anatomies by using a set of person-specific computational phantoms, is customizable to particular body counting facilities, applicable to any measurement setup, allows estimation of uncertainties with regard to the interindividual variation of the available phantoms, and is able to process information about the individual in form of anthropometric parameters to personalize calibration factors.

The framework is implemented in three components: a data model for general radiation protection scenarios, a software tool for computational body counter calibration, and an implementation of a statistical analysis method. This approach ensures continuity throughout the whole personalisation process, robust and algorithmically reproducible results, computational efficiency, and applicability to actual measurements and other types of body counting facilities.

20.1.1 Data model

The developed data model V2M Schema (chapter 11) was created based on the need for abstract, efficient and uniform processing of large, inhomogeneous and changing data from various sources. This refers to geometric models of phantoms and measurement equipment, nuclear decay data, media specifications, and source and tally locations. An abstraction layer in form of a taxonomy was added between geometry segments and media providing a basic approach for dynamic segment-medium mapping and universal segment identification through semantic annotation.

V2M Schema is an important step in data model standardization for radiation protection applications and, hopefully, gives a stimulus to future developments leading to efficient data exchange in the radiation protection community. The data model was inspired primarily by the Monte Carlo code MCNPX (Pelowitz 2007) and may require further generalization for other radiation transport codes. Interesting extensions of the data model are tallies comprising a binning with spatial resolution for computing isoflux surfaces, or a taxonomy for mapping between anatomical and biokinetic segments, which would be the basis for combining activity assessment and dose computation.

20.1.2 Software implementation

Voxel2MCNP (chapter 12) is a redesign of the corresponding software tool (Hegenbart 2009) developed at the former Institute for Radiation Research (ISF) and the Institute for Nuclear Waste Disposal (INE) at KIT. It is an established tool at the radiation protection group of the institute among researchers and students working with computational phantoms and MCNPX. The new version extends the original idea for modelling, simulation, and evaluation of radiation protection scenarios in an abstract and modular manner. It is fully integrated with V2M SCHEMA and provides data import and export capabilities for data formats and tools established in the radiation protection community, such as ENSDF (Tuli 2001), IMAGEJ (Ferreira and Rasband 2012), SIMPLEGEO (Theis et al. 2006), and MCNPX (Pelowitz 2007). There are also modules for scenario visualization and interactive geometry annotation, computation of anthropometric parameters, and automated and interactive detector positioning for body counting. Automatic evaluation of simulated pulse-height spectra is available in form of standard methods (Canberra 2006; International Organization for Standardization 2010a).

The new version of VOXEL2MCNP abstracts from the specifics of MCNPX and is designed to be extended to other Monte Carlo codes and applications beyond body counting as well as new data formats. This has been partially tested by modelling scenarios for computing organ absorbed dose fractions due to internal and external radiation sources (Pölz et al. 2013). An interesting future extension would be to perform the switch of the geometric representation of computational phantoms from voxel lattices to polygonal meshes in combination with a Monte Carlo code that is able process this format, which may alleviate current problems with high requirements on computer memory.

20.1.3 Statistical analysis method

The application of VOXEL2MCNP in combination with V2M SCHEMA to a specific body counter and a large phantom series enables the generation of huge data sets for sensitivity analysis. The implemented statistical analysis method (chapter 13) allows estimation of energy-dependent calibration factors for specific measurement setups based on anthropometric parameters. This is possible by combining established machine learning techniques for feature subset selection (Guyon and Elisseeff 2003) and kernel regression (Hastie, Tibshirani and Friedman 2009) with regard to structural risk minimization (Vapnik 1999).

The designed workflow has several free parameters and components. Sampling method, kernel method and performance measure can be optimized or replaced to be more suited for the particular application data in terms of accuracy and computing time. Depending on the available data, it would also be interesting to replace the selected wrapper method for feature subset selection with an embedded approach to increase computational efficiency.

20.2 Application

The main application of this work is the personalised calibration of the IVM body counter with the STEP framework. In addition, several sensitivity analyses were performed to quantify uncertainties beyond interindividual anatomical variation using components of STEP. These uncertainties are generally related to intraindividual variation, inhomogeneous source distributions, and accuracy in computational models and radiation transport simulation.

The IVM body counter was modelled (chapter 14) with all relevant components including detectors, their kinematics, and other equipment in the measurement chamber. The models were created with software tools and in data formats supported by VOXEL2MCNP, converted to V2M SCHEMA, and instantiated into calibration scenarios. Algorithmic interpretations of the outlined measurement setups, lungs-liver-knee and head (Marzocchi 2011), were defined from the perspective of a technician performing the positioning to ensure reproducibility. However, detector positioning remains a difficult task, because of its high sensitivity for low-energy photon emitters and especially for the liver, whose anatomical location varies considerably among the available computational phantoms.

Actually performing the modelling process on an example shows the advantages of the design approach. Existing models of the HPGe detectors (Marzocchi, Breustedt and Urban 2010) were recreated (Laubersheimer 2012) with SIMPLEGEO and are now primarily stored in this format. Concurrent model development and modelling of radiation transport scenarios using this approach is possible, since changes are automatically propagated on file import with VOXEL2MNCP. The data model provides a degree of abstraction and modularization that was not imaginable when modelling directly in MCNPX. The software supports beginners and experienced users of MCNPX regarding the simulation workflow and improves modelling and evaluation efficiency while reducing user mistakes.

20.2.1 Measurement reconstruction

Measurement reconstruction for physical phantoms is an important part of quality management of body counting facilities and validation of simulation models in international intercomparisons (Broggio et al. 2012; Gómez-Ros et al. 2008).

A study was performed with the JAERI torso phantom (Shirotani 1988) assessing the capability of reproducing measurements with computational methods (chapter 15). Lung sets with transuranic radionuclides such as ^{241}Am, ^{238}Pu, ^{235}U/^{238}U, and ^{232}Th were measured and reconstructed (Laubersheimer 2011; Laubersheimer 2012) with the help of VOXEL2MCNP, which allows easy import of radionuclide specifications, interactive detector positioning and analysis of pulse-height spectra.

The results show that simulation models in combination with radiation transports codes are a valid representation of physical measurements. There are, however, many small uncertainties regarding phantom models, detector positions and activity distributions that can add up to inconsistently large deviations of 15 % at maximum in the performed experiments with low-energy photon emitters.

20.2.2 ICRP-89 phantoms

Originally as a motivation for this work, four phantom series implementing the ICRP reference man specification (ICRP 2002) specifying body height, body mass, and organ masses were compared (chapter 16) with respect to body counting. The goal of this study was the analysis of uncertainties related to changes in body and organ shapes. These changes are inevitable since phantom developers use different imaging data or stylistic models. The considered phantom series were the ICRP series (ICRP 2009), the RPI series (Zhang et al. 2009), the UFPE series (Cassola et al. 2010), and the UFH series (Lee et al. 2007; Lee et al. 2010) each including an adult male and female phantom.

Considering a fixed photon energy of 59.5 keV as a representative for the low-energy range, relative standard deviation from the ICRP phantoms among all other phantoms is about 33 % for lungs, 43 % for liver, 15 % for knee, and 29 % for head. It turns out, that there are differences in the phantoms with respect to the specification related to the base data and development process. Keeping those changes in mind, it is still evident that body height, body mass, and organ masses define phantoms insufficiently with respect to body counting. They may, however, be better described by anthropometric parameters specifically selected for the measurement setup.

20.2.3 XCAT phantom series

The XCAT computational phantom series (Segars and Sturgeon 2010) was selected for the calibration of the IVM body counter with respect to varying anatomy (chapter 17). It is the only available series featuring person-specific phantoms, i.e. phantoms directly based on tomographic imaging data without further modification. This is quite contrary to the usual phantom development process in radiation protection inspired by the ICRP reference man specification (ICRP 2002), which leads to reference phantoms that are adapted to represent smaller groups of the population with various modelling techniques usually not based on additional imaging data (Cassola et al. 2011; Johnson et al. 2009; Na et al. 2010).

The series of 30 phantoms was applied to body counter calibration with VOXEL2MCNP for both standard measurement setups and 26 samples of energy in the energy range of 25 keV to 2 MeV. A reasonable compromise between accuracy and computational efficiency was taken. In addition, 19 anthropometric parameters, partially based on standard body measures for clothing sizes defined by EN 13402-1 (European Committee for Standardization 2001), were specified and computed using geometric algorithms with regard to actual measurement on persons during *in vivo* monitoring from the perspective of the performing technicians. Statistical analysis was applied to all samples for each pair of source structure and detector to determine subsets of sensitive anthropometric parameters, which were then used to estimate mean calibration curves. Meta optimization was performed to reduce overfitting to the available anatomies and to guarantee generalization of the estimator.

The results show largest deviations in the computed calibration values of the available phantoms for lungs, followed by liver, knee and head. They generally increase with a reduction in photon energy and can triple from 2 MeV to 25 keV. It was shown that these deviations can be reduced by creating estimators related to body circumferences close to the source structure. These are bust circumference for lungs, waist circumference for liver, thigh circumference for knee, and head circumference for head. Assessing these parameters leads to a reduction of relative standard deviation in determined calibration factors across the whole energy range of about 9 % for lungs, 17 % for liver, 11 % for knee, and 5 % for head compared to assessing no parameter. The reduction generally increases with photon energy. Whole-body measures, such as body mass and its derivatives, have a lower impact. The reduction for the legacy method (Mohr and Breustedt 2007) is 4 % for lungs, and 9 % for liver. Personalisation for head and knee cannot be performed with the associated torso phantom.

In conclusion, body circumferences at the vicinity of the source structure are positively correlated to local tissue thickness, which has a considerable impact on photon attenuation. Assessing these anthropometric parameters improves estimation of calibration factors compared to whole-body measures applied in standard personalisation methods (Doerfel, Heide and Sohlin 2006; Lynch 2011; Mohr and Breustedt 2007; Pierrat et al. 2007). These improvements directly impact the assessment of activities in organs and other source structures, which are the basis for calculation of organ absorbed and effective dose (ICRP 2007).

The results are optimal with respect to the applied phantom series and statistical analysis method. The phantom series was selected with care and the analysis method was constructed to generalize to other anatomies and reduce overfitting of the data. The main dependency on the success of the method is the number of available phantoms. Using more phantoms would provide a hold-out set for validation in the statistical model building process and possibly allow larger feature subsets. This can easily be achieved by applying the same workflows to additional phantoms of the XCAT series or other phantom series, and defining additional anthropometric parameters for specific measurement setups. The method can also be applied to different body counters by replacing the associated models and implemented measurement setups.

20.2.4 Inhomogeneous source distribution

There are additional uncertainties in body counting, which might even be larger than those related to interindividual anatomical variation. One of these is activity distribution in the source organ, which is not homogeneous in real applications — contrary to the general assumption in body counting. Radionuclide deposition in lungs is a case where the resulting distribution can be extremely inhomogeneous dependent on aerodynamic parameters (ICRP 1994).

An experimental study was conducted to quantify the uncertainty for lung counting with the IVM body counter (chapter 18). Measurements of the LLNL torso phantom (Griffith et al. 1987) with a perforated lung set that enables the creation of different source distributions were performed for the positron emitter ^{18}F.

In conclusion, the default assumption of a homogeneous source distribution produces high uncertainties with maximum changes in the order of a factor of 3 at 511 keV for the IVM body counter. Pelled et al. (2006) reported factors of 7 at 185 keV and 10 at 92 keV. This is comparable to the factors determined in this work considering the different photon energies.

20.2.5 Respiratory motion

The final study performed in this work regards intraindividual variation with respect to respiratory motion (chapter 19). While the impact of respiratory motion on body counting is certainly low, the availability of corresponding imaging data provided a good opportunity to familiarize with phantom development via image segmentation and to test the application of VOXEL2MCNP.

Respiratory-correlated computed tomography data sets (Guckenberger et al. 2007) were applied for the construction of a high-resolution four-dimensional torso phantom with eight time steps in the respiratory cycle (Schneider 2011a; Schneider 2011b). The results show a standard deviation of counting efficiency over a full respiratory cycle of about 1 % for tidal breathing since change in lung position at the superior part of the thorax is minimal. This is equivalent to a slight underestimation of counting efficiency for lung measurements when using a calibration phantom at full inspiration of tidal breathing. However, computed tomography data sets for phantom development are typically assessed at full inspiration of heavy breathing for better imaging of the thorax. Therefore, the observed effect on counting efficiency due to respiratory motion is expected to be larger considering those types of phantoms and a main issue in phantom development is the reliance on imaging data captured with these types of body deformation.

A V2M Schema files

The following listings are excerpts of several files that were used for the calibration of phantoms with VOXEL2MCNP (chapter 12). The files conform to the XML SCHEMA (XSD) (W3C 2004) format designed for the developed data model V2M SCHEMA (chapter 11). They describe scenarios (listing A.1), detectors (listing A.2), phantoms (listing A.3), materials (listing A.4), and source emissions (listing A.5).

Listing A.1: Scenario in V2M SCHEMA arranging all equipment resources in space. Each segment of an equipment can be associated with sources and tallies. The segments are identified by abstract terms of a taxonomy.

```xml
<scenario name="xcat050">
  <equipment-links>
    <equipment-link name="Phantom">
      <equipment-reference url="xcat050.v2m" name="xcat050"/>
      <source-links>
        <source-link name="Left Lung">
          <source-reference
            url="source-apeaks.v2m" name="Sample Peaks"/>
          <source-location>
            <term-id value="50"/> <!-- Left Lung -->
          </source-location>
        </source-link>
        <source-link name="Right Lung">
          <!-- ... -->
        </source-link>
        <!-- ... -->
      </source-links>
    </equipment-link>
    <equipment-link name="Detector: Left Lung">
      <equipment-reference
        url="hpge-detector3.v2m" name="HPGe Detector 3"/>
      <transformation>
        <rotation x="1.2947" y="0.1264" z="3.0398"/>
        <translation x="32.170" y="3.017" z="32.123"/>
      </transformation>
      <tally-links>
        <tally-link name="Left Lung">
          <tally-reference
            url="hpge-detector3.v2m" name="HPGe Detector 3"/>
          <tally-location>
            <term-id value="4"/> <!-- Crystal (Active) -->
          </tally-location>
        </tally-link>
      </tally-links>
    </equipment-link>
    <equipment-link name="Detector: Right Lung">
      <!-- ... -->
    </equipment-link>
    <!-- ... -->
  </equipment-links>
</scenario>
```

Listing A.2: Equipment and tally definition of a detector in V2M SCHEMA. The equipment section associates the geometry segments (e.g. parts) with abstract terms of a taxonomy for semantical interpretation. The actual definition of the surfaces is in an associated file. The tally section contains the number of detector channels, the energy range, and parameters describing energy resolution.

```
<equipment name="HPGe Detector 1">
  <description>
    HPGe detector 1 at the lower right position.
  </description>
  <segments>
    <taxonomy-reference
      url="detector-taxonomy.v2m" name="Detector"/>
    <materials-reference
      url="hpge-materials.v2m" name="HPGe"/>
    <geometry-reference
      url="hpge-detector1-geometry.v2m" name="HPGe Detector 1"/>
    <segments>
      &lt;segment id="1" name="Case" term-id="1"/&gt;
      &lt;segment id="2" name="Holder" term-id="2"/&gt;
      &lt;segment id="3" name="Window" term-id="3"/&gt;
      &lt;segment id="4" name="Crystal (Active)" term-id="4"/&gt;
      &lt;segment id="5" name="Crystal (Inactive)" term-id="5"/&gt;
      &lt;segment id="6" name="Vacuum" term-id="6"/&gt;
      <!-- ... -->
    </segments>
  </segments>
</equipment>
<tally name="HPGe Detector 1">
  <type value="pulse-height"/>
  <bins min="0" max="2.048" count="8192"/>
  <energy-resolution a="8.3195E-04" b="1.0985E-03" c="0"/>
</tally>
```

Listing A.3: Phantom definition in V2M SCHEMA. The geometry section describes the dimensions of the lattice structure and the voxel size. The compressed binary data is stored in an associated file. The equipment section associates geometry sements (e.g. organs and tissues) with abstract terms of a taxonomy for semantical interpretation.

```
<geometry name="xcat050">
  <cells>
    <repeated surface-id="1" id="1" name="lattice">
      <dimension x="648" y="263" z="1795"/>
      <binary-reference url="xcat050-binary.v2m" name="xcat050"/>
    </repeated>
  </cells>
  <surfaces>
    <surface type="box" id="1" name="voxel">
      <transformation>
        <scaling x="0.1" y="0.1" z="0.1"/>
      </transformation>
    </surface>
  </surfaces>
</geometry>
<equipment name="xcat050">
  <segments>
    <taxonomy-reference
      url="phantom-taxonomy.v2m" name="Anatomy"/>
    <materials-reference
      url="phantom-materials.v2m" name="ICRP/ICRU"/>
    <geometry-reference name="xcat050"/>
    <segments>
      &lt;segment id="-1046" name="Bronchi (115)" term-id="276"/&gt;
      <!-- ... -->
      &lt;segment id="815" name="Right Lung" term-id="181"/&gt;
      &lt;segment id="816" name="Left Lung" term-id="50"/&gt;
      <!-- ... -->
      &lt;segment id="819" name="Liver" term-id="87"/&gt;
      <!-- ... -->
      &lt;segment id="2266" name="Lesion" term-id="86"/&gt;
    </segments>
  </segments>
</equipment>
```

Listing A.4: Materials definition in V2M SCHEMA according to literature specification. Each material is described by density and elemental composition and is associated with abstract terms of a taxonomy for semantical interpretation.

```xml
<materials name="ICRP/ICRU">
  <description>
    General material specification for adult male and female
    phantoms. This is a combination of data from ICRU Report
    46, ICRP Publication 89, and ICRP Publication 110.
  </description>
  <taxonomy-reference url="phantom-taxonomy.v2m" name="Anatomy"/>
  <materials>
    <material id="301" name="Skeletal Muscle" term-id="315">
      <description>
        Skeletal muscle connecting joints along the skeleton.
      </description>
      <density value="1.050"/>
      <fractions type="mass-fractions">
        <fraction atomic-number="1" value="0.102"/>
        <fraction atomic-number="6" value="0.143"/>
        <fraction atomic-number="7" value="0.034"/>
        <fraction atomic-number="8" value="0.710"/>
        <fraction atomic-number="11" value="0.001"/>
        <fraction atomic-number="15" value="0.002"/>
        <fraction atomic-number="16" value="0.003"/>
        <fraction atomic-number="17" value="0.001"/>
        <fraction atomic-number="19" value="0.004"/>
      </fractions>
    </material>
    <!-- ... -->
  </materials>
</materials>
```

Listing A.5: Source definition in V2M SCHEMA adjusted to the operational detector range and typical calibration curves. Each source emission specifies energy, particle type, and intensity of the emission.

```xml
<source name="Sample Peaks">
  <description>
    A collection of 26 artificial photon emission peaks covering
    a large range of energies optimized for the HPGe binning
    (each peak is centred to a channel).
  </description>
  <emissions>
    <emission energy="0.010875" intensity="1" type="gamma"/>
    <!-- ... -->
    <emission energy="0.104875" intensity="1" type="gamma"/>
    <emission energy="0.129875" intensity="1" type="gamma"/>
    <emission energy="0.154875" intensity="1" type="gamma"/>
    <!-- ... -->
    <emission energy="2.047875" intensity="1" type="gamma"/>
  </emissions>
</source>
```

B Anthropometric parameters

The considered anthropometric parameters for sensitivity analysis of the XCAT series (section 17.2) are grouped into three categories: masses and derivatives (table B.1), lengths, breadths and distances (table B.2), and circumferences (table B.3). Details regarding physical measurement *(physical)* and algorithmic computation *(virtual)* are indicated by according keywords. Additionally, average photon transmission (section 12.4) was computed for each source organ and primary detector for each simulated photon energy.

Measure	Description
Body mass	Total body mass. *Physical:* The person is measured with a balance in an overall without shoes. *Virtual:* Computed from the geometrical volumes and associated material densities.
Mass/Height	Ratio of body mass (in kg) and height (in m).
Body mass index (BMI)	Ratio of body mass (in kg) and squared height (in m^2).
Body fat percentage	Ratio of adipose tissue mass and body mass. *Physical:* Bioelectrical impedance analysis with a body fat meter. *Virtual:* Identification of segments that are annotated as adipose tissue.

Table B.1: Computation and measurement of body volumes, masses and derivatives.

Measure	Description
Body height	Total body length between foot soles and crest of the head in standing position. *Physical:* The person is measured with a height gauge standing with closed feet and without shoes. *Virtual:* Phantoms are usually based on tomographic imaging data taken in supine position. The body length is corrected for angled feet.
Inner leg length	Distance between distal end of the pelvis and distal end of the feet in standing position. *Physical:* The person is measured with a tape measure standing with legs slightly apart. *Virtual:* Similar to body height, the leg length is corrected for angled feet.
Arm length	Distance between the lateral end of the acromion process across the elbow to the distal end of the ulna on the right side of the body. *Physical:* The person is standing with the right arm bent at 90° and the fist placed on the hip. *Virtual:* The distance is computed on the dorsal skin surface.
Shoulder breadth	Distance between the lateral ends of the acromion processes. *Physical:* The person is measured with a tape measure across the back. *Virtual:* The direct distance of the acromion processes.

Table B.2: Computation and measurement of body lengths, breadths and distances.

Measure	Description
All	*Physical:* Circumferences are measured with a tape measure with the person either standing or sitting. *Virtual:* Geometric algorithms (section 12.4) are applied.
Head c.	Largest circumference of the head slightly above the eyebrows and the ears.
Chest c.	Circumference across the chest (slightly below the sternal angle) and scapulae, and below the armpits. Slightly angled with respect to the transversal plane.
Bust c.	Largest circumference across the most prominent part of the bust and scapulae, and below the armpits. Slightly above the inferior angle of the scapulae.
Underbust c.	Circumference directly under the bust across the infrasternal notch and the inferior angle of the scapulae.
Waist c.	Circumference between the iliac crest and the lower ribs.
Hip c.	Largest circumference at the buttocks.
Thigh c.	Largest circumference of the leg between hip and knee.
Calf c.	Largest circumference of the leg between knee and ankle.
Upper arm c.	Largest circumference of the arm between shoulder and elbow.
Lower arm c.	Largest circumference of the arm between elbow and wrist.

Table B.3: Computation and measurement of body circumferences based on anatomic landmarks.

Related publications

If you are unable to find any of the following publications, feel free to request a full text version at stefan.poelz@gmail.com.

Peer-reviewed articles

HEGENBART, L., S. PÖLZ, A. BENZLER and M. URBAN (2012). 'Voxel2MCNP: Software for handling voxel models for Monte Carlo radiation transport calculations'. In: *Health Physics* **102** (2): 221–229. DOI: 10.1097/HP.0b013e3182321cdb.

PÖLZ, S., S. LAUBERSHEIMER, J. S. EBERHARDT, M. A. HARRENDORF, T. KECK, A. BENZLER and B. BREUSTEDT (2013). 'Voxel2MCNP: A framework for modeling, simulation and evaluation of radiation transport scenarios for Monte Carlo codes'. In: *Physics in Medicine and Biology* **58** (16): 5381–5400. DOI: 10.1088/0031-9155/58/16/5381.

Conference contributions

PÖLZ, S. (2011). 'Comparison of anthropomorphic reference phantoms for numerical efficiency calibration'. In: *The 3rd International Workshop on Computational Phantoms for Radiation Protection, Imaging and Radiotherapy, 8-9 August 2011*. Beijing, China.

PÖLZ, S. (2012). 'Numerical efficiency calibration of partial body counters using anthropomorphic phantoms'. In: *79. Sitzung des Arbeitskreises Inkorporationsüberwachung, 17 September 2012*. Karlsruhe, Germany.

PÖLZ, S. and B. BREUSTEDT (2013a). 'Calibration of partial body counters using Voxel2MCNP and the XCAT phantom series'. In: *4th International Workshop on Computational Phantoms for Radiation Protection, Imaging and Radiotherapy, 20-22 May 2013*. Zurich, Switzerland.

PÖLZ, S., M. URBAN and H. GECKEIS (2012). 'Numerische Effizienzkalibrierung von Teilkörperzählern mit Hilfe von anthropomorphen Phantomen'. In: *Jahrestagung Kerntechnik, 22-24 May 2012*. Stuttgart, Germany.

PÖLZ, S., T. SCHNEIDER, L. HEGENBART, M. URBAN and H. GECKEIS (2012). 'Quantification of the effect of respiratory motion on efficiency calibration for internal dosimetry'. In: *The 13th International Congress of the International Radiation Protection Association, 13-18 May 2012*. Glasgow, Scotland.

Reports

INSTITUT FÜR STRAHLENFORSCHUNG (2011). *Jahresbericht 2010*. Tech. rep. KIT-SR 7587. Karlsruhe, Germany: Karlsruhe Institute of Technology: 55–58,60–61,66–69,69–74.

INSTITUTE FOR NUCLEAR WASTE DISPOSAL (2012). *Annual Report 2011*. Tech. rep. KIT-SR 7617. Karlsruhe, Germany: Karlsruhe Institute of Technology: 81–85.

PÖLZ, S. and B. BREUSTEDT (2011a). 'Zähleffizienzkalibrierung von in vivo Messsystemen mit probanden-adaptierten anthropomorphen Modellen'. In: *Workshop Strahlung und Umwelt II: Radionuklide in der Umwelt, ihr Transport in Nahrungsketten zum und im Menschen, 15 March 2011*. Eggenstein-Leopoldshafen, Germany.

PÖLZ, S. and B. BREUSTEDT (2011b). 'Zähleffizienzkalibrierung von in vivo Messsystemen mit probanden-adaptierten anthropomorphen Modellen'. In: *Workshop Strahlung und Umwelt II: Radionuklide in der Umwelt, ihr Transport in Nahrungsketten zum und im Menschen, 15 September 2011*. Rheinbach, Germany.

PÖLZ, S. and B. BREUSTEDT (2012a). 'Zähleffizienzkalibrierung von in vivo Messsystemen mit probanden-adaptierten anthropomorphen Modellen'. In: *Workshop Strahlung und Umwelt II: Radionuklide in der Umwelt, ihr Transport in Nahrungsketten zum und im Menschen, 28-29 February 2012*. Neuherberg, Germany.

PÖLZ, S. and B. BREUSTEDT (2012b). 'Zähleffizienzkalibrierung von in vivo Messsystemen mit probanden-adaptierten anthropomorphen Modellen'. In: *Workshop Strahlung und Umwelt II: Radionuklide in der Umwelt, ihr Transport in Nahrungsketten zum und im Menschen, 8-9 October 2012*. Jena, Germany.

PÖLZ, S. and B. BREUSTEDT (2013b). 'Zähleffizienzkalibrierung von in vivo Messsystemen mit probanden-adaptierten anthropomorphen Modellen'. In:

Workshop Strahlung und Umwelt II: Radionuklide in der Umwelt, ihr Transport in Nahrungsketten zum und im Menschen, 20-21 June 2013. Neuherberg, Germany.

PROJEKTTRÄGER KARLSRUHE WASSERTECHNOLOGIE UND ENTSORGUNG (2011a). *Nukleare Sicherheitsforschung.* Tech. rep. PTE-N 2. Karlsruhe, Germany: Karlsruhe Institute of Technology: 164–165.

PROJEKTTRÄGER KARLSRUHE WASSERTECHNOLOGIE UND ENTSORGUNG (2011b). *Nukleare Sicherheitsforschung.* Tech. rep. PTE-N 3. Karlsruhe, Germany: Karlsruhe Institute of Technology: 160–161.

PROJEKTTRÄGER KARLSRUHE WASSERTECHNOLOGIE UND ENTSORGUNG (2012a). *Nukleare Sicherheitsforschung.* Tech. rep. PTE-N 4. Karlsruhe, Germany: Karlsruhe Institute of Technology: 154–155.

PROJEKTTRÄGER KARLSRUHE WASSERTECHNOLOGIE UND ENTSORGUNG (2012b). *Nukleare Sicherheitsforschung.* Tech. rep. PTE-N 5. Karlsruhe, Germany: Karlsruhe Institute of Technology: 148–149.

PROJEKTTRÄGER KARLSRUHE WASSERTECHNOLOGIE UND ENTSORGUNG (2013a). *Nukleare Sicherheitsforschung.* Tech. rep. PTE-N 6. Karlsruhe, Germany: Karlsruhe Institute of Technology: 168–169.

PROJEKTTRÄGER KARLSRUHE WASSERTECHNOLOGIE UND ENTSORGUNG (2013b). *Nukleare Sicherheitsforschung.* Tech. rep. PTE-N 7. Karlsruhe, Germany: Karlsruhe Institute of Technology: 162–163.

Supervised works

SCHNEIDER, T. (2011a). 'Auswirkung der Atembewegung auf die Zähleffizienz von in vivo Messsystemen bei Lungenteilkörpermessungen'. Bachelor thesis. Duale Hochschule Baden-Württemberg Karlsruhe.

SCHNEIDER, T. (2011b). 'Erstellung von Voxelmodellen zur Bestimmung der Atembewegungsabhängigkeit der Zähleffizienz bei in vivo Inkorporationsmessungen'. Student thesis. Duale Hochschule Baden-Württemberg Karlsruhe.

References

ABDI, H. and L. J. WILLIAMS (2010). 'Principal component analysis'. In: *Wiley Interdisciplinary Reviews: Computational Statistics* **2** (4): 433–459. DOI: 10.1002/wics.101.

ABLE SOFTWARE CORP (2013). *3D-Doctor*. Lexington, MA. URL: www.ablesw.com/3d-doctor.

AGOSTINELLI, S., J. ALLISON, K. AMAKO, J. APOSTOLAKIS, H. ARAUJO, P. ARCE, M. ASAI, D. AXEN, S. BANERJEE, G. BARRAND, F. BEHNER, L. BELLAGAMBA, J. BOUDREAU, L. BROGLIA, A. BRUNENGO, H. BURKHARDT, S. CHAUVIE, J. CHUMA, R. CHYTRACEK, G. COOPERMAN, G. COSMO, P. DEGTYARENKO, A. DELL'ACQUA, G. DEPAOLA, D. DIETRICH, R. ENAMI, A. FELICIELLO, C. FERGUSON, H. FESEFELDT, G. FOLGER, F. FOPPIANO, A. FORTI, S. GARELLI, S. GIANI, R. GIANNITRAPANI, D. GIBIN, J. J. GÓMEZ CADENAS, I. GONZÁLEZ, G. GRACIA ABRIL, G. GREENIAUS, W. GREINER, V. GRICHINE, A. GROSSHEIM, S. GUATELLI, P. GUMPLINGER, R. HAMATSU, K. HASHIMOTO, H. HASUI, A. HEIKKINEN, A. HOWARD, V. IVANCHENKO, A. JOHNSON, F. W. JONES, J. KALLENBACH, N. KANAYA, M. KAWABATA, Y. KAWABATA, M. KAWAGUTI, S. KELNER, P. KENT, A. KIMURA, T. KODAMA, R. KOKOULIN, M. KOSSOV, H. KURASHIGE, E. LAMANNA, T. LAMPÉN, V. LARA, V. LEFEBURE, F. LEI, M. LIENDL, W. LOCKMAN, F. LONGO, S. MAGNI, M. MAIRE, E. MEDERNACH, K. MINAMIMOTO, P. MORA DE FREITAS, Y. MORITA, K. MURAKAMI, M. NAGAMATU, R. NARTALLO, P. NIEMINEN, T. NISHIMURA, K. OHTSUBO, M. OKAMURA, S. O'NEALE, Y. OOHATA, K. PAECH, J. PERL, A. PFEIFFER, M. G. PIA, F. RANJARD, A. RYBIN, S. SADILOV, E. DI SALVO, G. SANTIN, T. SASAKI, N. SAVVAS, Y. SAWADA, S. SCHERER, S. SEI, V. SIROTENKO, D. SMITH, N. STARKOV, H. STOECKER, J. SULKIMO, M. TAKAHATA, S. TANAKA, E. TCHERNIAEV, E. SAFAI TEHRANI, M. TROPEANO, P. TRUSCOTT, H. UNO, L. URBAN, P. URBAN, M. VERDERI, A. WALKDEN, W. WANDER, H. WEBER, J. P. WELLISCH, T. WENAUS, D. C. WILLIAMS, D. WRIGHT, T. YAMADA, H. YOSHIDA and D. ZSCHIESCHE (2003). 'GEANT4 — A simulation toolkit'. In: *Nuclear Instruments and Methods in Physics Research Section A: Accelerators, Spectrometers, Detectors and Associated Equipment* **506** (3): 250–303. DOI: 10.1016/S0168-9002(03)01368-8.

ARLOT, S. and A. CELISSE (2010). 'A survey of cross-validation procedures for model selection'. In: *Statistics Surveys* **4**: 40–79. DOI: 10.1214/09-SS054.

ATTIX, F. H. (1991). *Introduction to radiological physics and radiation dosimetry.* 1st ed. Weinheim, GER: WILEY-VCH Verlag GmbH & Co. KGaA.

BALASHAZY, I., W. HOFMANN and A. FARKAS (2002). 'Numerical modeling of deposition of inhaled particles in central human airways'. In: *Annals of Occupational Hygiene* **46** (Suppl. 1): 353–357. DOI: 10.1093/annhyg/46.suppl_1.353.

BEKKERMAN, R., R. EL-YANIV, N. TISHBY and Y. WINTER (2003). 'Distributional word clusters vs. words for text categorization'. In: *Journal of Machine Learning Research* **3**: 1183–1208. URL: http://dl.acm.org/citation.cfm?id=944969.

BERGER, M. J., J. S. COURSEY, M. A. ZUCKER and J. CHANG (2005). *Stopping-power and range tables for electrons, protons, and helium ions.* Tech. rep. NISTIR 4999. Gaithersburg, MD: National Institute of Standards and Technology. URL: www.nist.gov/pml/data/star.

BERGER, M. J., J. H. HUBBELL, S. M. SELTZER, J. CHANG, J. S. COURSEY, R. SUKUMAR, D. S. ZUCKER and K. OLSEN (2010). *XCOM: Photon cross section database.* Tech. rep. NBSIR 87-3597. Gaithersburg, MD: National Institute of Standards and Technology. URL: www.nist.gov/pml/data/xcom.

BERGMANN, R., W. HOFMANN and L. KOBLINGER (1997). 'The effect of ventilation inhomogeneities on aerosol deposition and bolus dipersion'. In: *Annals of Occupational Hygiene* **41** (Inhaled Particles VIII): 543–547. DOI: 10.1093/annhyg/41.inhaled_particles_VIII.543.

BIRD, A. J. and T. C. FRY (2013). 'Visual Workshop 2: A model viewer, editor and results display package for the ANSWERS shielding and criticality codes'. In: *Progress in Nuclear Science and Technology.*

BOLCH, W., C. LEE, M. WAYSON and P. JOHNSON (2010). 'Hybrid computational phantoms for medical dose reconstruction'. In: *Radiation and Environmental Biophysics* **49** (2): 155–168. DOI: 10.1007/s00411-009-0260-x.

BREIMAN, L. (1994). *Bagging predictors.* Tech. rep. 421. Berkeley, CA: University of California at Berkeley. DOI: 10.1007/BF00058655.

BREIMAN, L. (1996). 'Stacked regressions'. In: *Machine Learning* **24** (1): 49–64. DOI: 10.1007/BF00117832.

BRITTON, R., J. BURNETT, A. DAVIES and P. H. REGAN (2012). 'Determining the efficiency of a broad-energy HPGe detector using Monte Carlo simulations'. In: *Journal of Radioanalytical and Nuclear Chemistry* **295** (3): 2035–2041. DOI: 10.1007/s10967-012-2203-2.

BROGGIO, D., J. BENTO, M. CALDEIRA, E. CARDENAS-MENDEZ, J. FARAH, T. FONSECA, C. KONVALINKA, L. LIU, B. PEREZ, K. CAPELLO, P. COWAN, J.-A. CRUZATE, L. FREIRE, J.-M. GÓMEZ-ROS, S. GOSSIO, B. HEIDE, J. HUIKARI, J. HUNT, S. KINASE, G. H. KRAMER, O. KURIHARA, A. KYRIELEIS, A.-L. LEBACQ, D. LEONE, C. LI, J. LI, L.-C. MIHAILESCU, M. MORALEDA, J.-F. NAVARRO, C. OLIVEIRA, N. PUERTA, U. REICHELT, C. SIMÕES, D. SOMMER, M. TAKAHASHI, P. TELES, F. VANHAVERE, T. VRBA, D. FRANCK, G. GUALDRINI and M.-A. LOPEZ (2012). 'Monte Carlo modelling for the in vivo lung monitoring of enriched uranium: Results of an international comparison'. In: *Radiation Measurements* **47** (7): 492–500. DOI: 10.1016/j.radmeas.2012.04.020.

CANBERRA (2006). *Genie 2000 spectroscopy software — Customization tools.* Tech. rep. 9233653F. Meriden, CT: Canberra Industries, Inc.

CANBERRA (2008). *Germanium detectors.* Tech. rep. C36151. Meriden, CT: Canberra Industries, Inc.

CANBERRA (2009). *Genie 2000 gamma analysis software.* Tech. rep. C37593. Meriden, CT: Canberra Industries, Inc.

CANBERRA (2012). *Cryo-pulse 5 plus electrically refrigerated cryostat.* Tech. rep. C39805. Meriden, CT: Canberra Industries, Inc.

CANBERRA (2013). *Extended range coaxial Ge detectors (XtRa).* Tech. rep. C40024. Meriden, CT: Canberra Industries, Inc.

CARROLL, R. J., D. RUPPERT, L. A. STEFANSKI and C. M. CRAINICEANU (2006). *Measurement error in nonlinear models: A modern perspective (Monographs on statistics and applied probability).* 2nd ed. Boca Raton, FL: Chapman & Hall/CRC.

CASSOLA, V. F., V. J. D. M. LIMA, R. KRAMER and H. J. KHOURY (2010). 'FASH and MASH: Female and male adult human phantoms based on polygon mesh surfaces: I. Development of the anatomy'. In: *Physics in Medicine and Biology* **55** (1): 133–162. DOI: 10.1088/0031-9155/55/1/009.

CASSOLA, V. F., F. M. MILIAN, R. KRAMER, C. A. B. DE OLIVEIRA LIRA and H. J. KHOURY (2011). 'Standing adult human phantoms based on 10th, 50th and 90th mass and height percentiles of male and female Caucasian populations'. In: *Physics in Medicine and Biology* **56** (13): 3749–3772. DOI: 10.1088/0031-9155/56/13/002.

CHARLES, M. W. and J. D. HARRISON (2007). 'Hot particle dosimetry and radiobiology — Past and present'. In: *Journal of Radiological Protection: Official Journal of the Society for Radiological Protection* **27** (3A): A97–109. DOI: 10.1088/0952-4746/27/3A/S11.

CHIAVASSA, S., M. BARDIÈS, F. GUIRAUD-VITAUX, D. BRUEL, J.-R. JOURDAIN, D. FRANCK and I. AUBINEAU-LANIÈCE (2005). 'OEDIPE: A personalized dosimetric tool associating voxel-based models with MCNPX'. In: *Cancer Biotherapy & Radiopharmaceuticals* **20** (3): 325–332. DOI: 10.1089/cbr.2005.20.325.

CLEVELAND, W. S. (1979). 'Robust locally weighted regression and smoothing scatterplots'. In: *Journal of the American Statistical Association* **74** (368): 829–836. DOI: 10.1080/01621459.1979.10481038.

CLEVELAND, W. S. and S. J. DEVLIN (1988). 'Locally weighted regression: An approach to regression analysis by local fitting'. In: *Journal of the American Statistical Association* **83** (403): 596–610. URL: www.jstor.org/stable/2289282.

COURAGEOT, E., R. SAYAH and C. HUET (2010). 'Development of modified voxel phantoms for the numerical dosimetric reconstruction of radiological accidents involving external sources: implementation in SESAME tool'. In: *Physics in Medicine and Biology* **55** (9): N231–241. DOI: 10.1088/0031-9155/55/9/N02.

COWAN, P., G. DOBSON and J. MARTIN (2013). 'Release of MCBEND 11'. In: *Progress in Nuclear Science and Technology*.

DEPARTMENT OF HEALTH AND HUMAN SERVICES and NATIONAL CENTER FOR HEALTH STATISTICS (1996). 'NHANES III examination data file'. In: *The Third National Health and Nutrition Examination Survey 1988-1994* **11** (1A). URL: www.cdc.gov/nchs/nhanes/nh3data.htm.

DHILLON, I. S., S. MALLELA and R. KUMAR (2003). 'A divisive information theoretic feature clustering algorithm for text classification'. In: *Journal of Machine Learning Research* **3**: 1265–1287. URL: http://dl.acm.org/citation.cfm?id=944973.

DIMBYLOW, P. and R. FINDLAY (2010). 'The effects of body posture, anatomy, age and pregnancy on the calculation of induced current densities at 50 Hz'. In: *Radiation Protection Dosimetry* **139** (4): 532–538. DOI: 10.1093/rpd/ncp285.

DOERFEL, H., B. HEIDE and M. SOHLIN (2006). *Entwicklung eines Verfahrens zur numerischen Kalibrierung von Teilkörperzählern*. Tech. rep. FZKA 7238.

DOUGHERTY, E. R. (1992). *An introduction to morphological image processing*. SPIE Optical Engineering Press.

DRUCKER, H., C. J. C. BURGES, L. KAUFMAN, A. SMOLA and V. VAPNIK (1997). 'Support vector regression machines'. In: *Advances in Neural Information*

Processing Systems 9. Ed. by M. C. MOZER, M. I. JORDAN and T. PETSCHE. Vol. 1. MIT Press: 155–161. DOI: `10.1.1.21.5909`.

ELANIQUE, A., O. MARZOCCHI, D. LEONE, L. HEGENBART, B. BREUSTEDT and L. OUFNI (2012). 'Dead layer thickness characterization of an HPGe detector by measurements and Monte Carlo simulations'. In: *Applied Radiation and Isotopes: Including Data, Instrumentation and Methods for Use in Agriculture, Industry and Medicine* **70** (3): 538–542. DOI: `10.1016/j.apradiso.2011.11.014`.

EUROPEAN COMMITTEE FOR STANDARDIZATION (2001). *Size designation of clothes — Part 1: Terms, definitions and body measurement procedure*. Tech. rep. EN 13402-1:2001 (ISO 3635:1981 modified). Brussels: European Committee for Standardization.

FARAH, J., D. BROGGIO and D. FRANCK (2010). 'Creation and use of adjustable 3D phantoms: application for the lung monitoring of female workers'. In: *Health Physics* **99** (5): 649–661. DOI: `10.1097/HP.0b013e3181dc4f58`.

FARAH, J., D. BROGGIO and D. FRANCK (2011a). 'Efficient calculation of in vivo efficiency curves using variance reduction techniques'. In: *Progress in Nuclear Science and Technology* **2**: 247–252.

FARAH, J., D. BROGGIO and D. FRANCK (2011b). 'Examples of mesh and NURBS modelling for in vivo lung counting studies'. In: *Radiation Protection Dosimetry* **144** (1-4): 344–348. DOI: `10.1093/rpd/ncq313`.

FARFÁN, E. B., E. Y. HAN, W. E. BOLCH, C. HUH, T. E. HUSTON and W. E. BOLCH (2004). 'A revised stylized model of the adult extrathoracic and thoracic airways for use with the ICRP-66 human respiratory tract model'. In: *Health Physics* **86** (4): 337–352. URL: `www.ncbi.nlm.nih.gov/pubmed/15057054`.

FASSÒ, A., A. FERRARI, S. ROESLER, J. RANFT, P. R. SALA, G. BATTISTONI, M. CAMPANELLA, F. CERUTTI, L. D. BIAGGI, E. GADIOLI, M. V. GARZELLI, F. BALLARINI, A. OTTOLENGHI, D. SCANNICCHIO, M. CARBONI, M. PELLICCIONI, R. VILLARI, V. ANDERSEN, A. EMPL, K. LEE, L. PINSKY, T. N. WILSON and N. ZAPP (2003). 'The FLUKA code: Present applications and future developments'. In: *2003 Conference for Computing in High Energy and Nuclear Physics*. La Jolla, CA. URL: `http://arxiv.org/abs/physics/0306162`.

FEDERATIVE COMMITTEE ON ANATOMICAL TERMINOLOGY (1998). *Terminologia anatomica: International anatomical terminology*. Ed. by I. WHITMORE. 2nd ed. New York: Thieme.

FEDERATIVE INTERNATIONAL COMMITTEE ON ANATOMICAL TERMINOLOGY (2007). *Terminologia histologica: International terms for human cytology and histology*. Baltimore, MD: Lippincott Williams & Wilkins.

FERREIRA, T. and W. RASBAND (2012). *ImageJ user guide*. Tech. rep. IJ 1.46r. Bethesda, MD: National Institutes of Health.

FIRESTONE, R. B. and L. P. EKSTRÖM (2004). *WWW table of radioactive isotopes, database version 2.1, January 2004*. URL: http://ie.lbl.gov/toi.

GLOVER, F. and C. MCMILLAN (1986). 'The general employee scheduling problem: An integration of MS and AI'. In: *Computers & Operations Research* **13** (5): 563–573. DOI: 10.1016/0305-0548(86)90050-X.

GOLDBERG, D. E. (1989). *Genetic algorithms in search, optimization and machine learning*. 1st ed. Boston, MA: Addison-Wesley Longman Publishing Co., Inc.

GÓMEZ-ROS, J. M., L. DE CARLAN, D. FRANCK, G. GUALDRINI, M. LIS, M. A. LÓPEZ, M. MORALEDA, M. ZANKL, A. BADAL, K. CAPELLO, P. COWAN, P. FERRARI, B. HEIDE, J. HENNIGER, V. HOOLEY, J. HUNT, S. KINASE, G. H. KRAMER, D. LÖHNERT, S. LUCAS, V. NUTTENS, L. W. PACKER, U. REICHELT, T. VRBA, J. SEMPAU and B. ZHANG (2008). 'Monte Carlo modelling of Germanium detectors for the measurement of low energy photons in internal dosimetry: Results of an international comparison'. In: *Radiation Measurements* **43** (2-6): 510–515. DOI: 10.1016/j.radmeas.2007.12.023.

GRAHAM, R. L. (1972). 'An efficient algorithm for determining the convex hull of a finite planar set'. In: *Information Processing Letters* **1**: 132–133.

GRAY, P. W. and A. AHMAD (1985). 'Linear classes of Ge(Li) detector efficiency functions'. In: *Nuclear Instruments and Methods in Physics Research Section A: Accelerators, Spectrometers, Detectors and Associated Equipment* **237** (3): 577–589. DOI: 10.1016/0168-9002(85)91069-1.

GRIFFITH, R. V., A. L. ANDERSON, P. N. DEAN, J. C. FISHER and C. W. SUNDBECK (1987). 'Tissue-equivalent torso phantom for calibration of transuranic-nuclide counting facilities'. In: *Proceedings of the Department of Energy Workshop on Radiobioassay and Internal Dosimetry*. Albuquerque, NM: 293–314.

GROTHENDIECK, G. (2010). *nls2: Non-linear regression with brute force*. Tech. rep. URL: http://cran.r-project.org/package=nls2.

GUCKENBERGER, M., J. WILBERT, T. KRIEGER, A. RICHTER, K. BAIER, J. MEYER and M. FLENTJE (2007). 'Four-dimensional treatment planning for stereotactic body radiotherapy'. In: *International Journal of Radiation Oncology, Biology, Physics* **69** (1): 276–285. DOI: 10.1016/j.ijrobp.2007.04.074.

GUCKENBERGER, M., R. A. SWEENEY, J. WILBERT, T. KRIEGER, A. RICHTER, K. BAIER, G. MUELLER, O. SAUER and M. FLENTJE (2008). 'Image-guided radiotherapy for liver cancer using respiratory-correlated computed tomography and cone-beam computed tomography'. In: *International Journal of Radiation Oncology, Biology, Physics* **71** (1): 297–304. DOI: 10.1016/j.ijrobp.2008.01.005.

GÜN, H. (2010). 'Bestimmung der Brustwandstärke als Kalibrierparameter für Teilkörpermessungen'. Diploma thesis. Karlsruhe Institute of Technology.

GUNDIMADA, S., V. K. ASARI and N. GUDUR (2010). 'Face recognition in multi-sensor images based on a novel modular feature selection technique'. In: *Information Fusion* **11** (2): 124–132. DOI: 10.1016/j.inffus.2009.05.002.

GUYON, I. and A. ELISSEEFF (2003). 'An introduction to variable and feature selection'. In: *The Journal of Machine Learning Research* **3**: 1157–1182. URL: http://dl.acm.org/citation.cfm?id=944968.

GUYON, I., J. WESTON, S. BARNHILL and V. VAPNIK (2002). 'Gene selection for cancer classification using support vector machines'. In: *Machine Learning* **46** (1-3): 389–422. DOI: 10.1023/A:1012487302797.

HART, J. C. (1996). 'Sphere tracing: A geometric method for the antialiased ray tracing of implicit surfaces'. In: *The Visual Computer* **12**: 527–545.

HASTIE, T., R. TIBSHIRANI and J. J. H. FRIEDMAN (2009). *The elements of statistical learning — Data mining, inference, and prediction.* 2nd ed. Springer.

HEGENBART, L., Y. H. NA, J. Y. ZHANG, M. URBAN and X. G. XU (2008). 'A Monte Carlo study of lung counting efficiency for female workers of different breast sizes using deformable phantoms'. In: *Radiation Protection Dosimetry* **53** (19): 5527–5538. DOI: 10.1088/0031-9155/53/19/017.

HEGENBART, L., O. MARZOCCHI, B. BREUSTEDT and M. URBAN (2009). 'Validation of a Monte Carlo efficiency calibration procedure for a partial body counter system with a voxel model of the LLNL torso phantom'. In: *Radiation Protection Dosimetry* **133** (3): 158–164. DOI: 10.1093/rpd/ncp030.

HEGENBART, L. (2009). 'Numerical efficiency calibration of in vivo measurement systems'. PhD thesis. Karlsruhe Institute of Technology. URL: http://digbib.ubka.uni-karlsruhe.de/volltexte/1000017470.

HEGENBART, L. and B. BREUSTEDT (2011). 'A new position recording system for the partial-body counter at KIT'. In: *Radiation Protection Dosimetry* **144** (1-4): 389–392. DOI: 10.1093/rpd/ncq416.

HEGENBART, L., H. GÜN and M. ZANKL (2010). 'Determination of chest wall thickness of anthropometric voxel models'. In: Helsinki, Finland: 1–9.

HEGENBART, L. and H. GÜN (2010). 'Determination of chest wall thickness of anthropometric voxel models (Poster)'. In: *Proceedings of the Third European IRPA Congress.* Helsinki, Finland.

HEGENBART, L., S. PÖLZ, A. BENZLER and M. URBAN (2012). 'Voxel2MCNP: Software for handling voxel models for Monte Carlo radiation transport calculations'. In: *Health Physics* **102** (2): 221–229. DOI: 10.1097/HP.0b013e3182321cdb.

HENRIET, J., P.-E. LENI, R. LAURENT, A. ROXIN, B. CHEBEL-MORELLO, M. SALOMON, J. FARAH, D. BROGGIO, D. FRANCK and L. MAKOVICKA (2012). 'Adapting numerical representations of lung contours using case-based reasoning and artificial neural networks'. In: *Case-Based Reasoning Research and Development.* Ed. by B. D. AGUDO and I. WATSON. Lyon, France: Springer-Verlag Berlin Heidelberg: 137–151.

HUBBELL, J. H. and S. M. SELTZER (2004). *Tables of X-ray mass attenuation coefficients and mass energy-absorption coefficients from 1 keV to 20 MeV for elements Z = 1 to 92 and 48 additional substances of dosimetric interest.* Tech. rep. NISTIR 5632. Gaithersburg, MD: National Institute of Standards and Technology. URL: www.nist.gov/pml/data/xraycoef.

HUET, C., A. LEMOSQUET, I. CLAIRAND, J. B. RIOUAL, D. FRANCK, L DE CARLAN, I. AUBINEAU-LANIÈCE and J. F. BOTTOLLIER-DEPOIS (2009). 'SESAME: A software tool for the numerical dosimetric reconstruction of radiological accidents involving external sources and its application to the accident in Chile in December 2005'. In: *Health Physics* **96** (1): 76–83. DOI: 10.1097/01.HP.0000327661.90794.0f.

HUNT, J. G., B. M. DANTAS, M. C. LOURENÇO and A. M. G. AZEREDO (2003). 'Voxel phantoms and Monte Carlo methods applied to in vivo measurements for simultaneous 241-Am contamination in four body regions'. In: *Radiation Protection Dosimetry* **105** (1-4): 549–552.

IAEA (1996). 'Direct methods for measuring radionuclides in the human body'. In: *Safety Series* (114).

IAEA (2004). 'Methods for assessing occupational radiation doses due to intakes of radionuclides'. In: *Safety Reports Series* (37).

IAEA and ILO (1999). 'Assessment of occupational exposure due to intakes of radionuclides'. In: *IAEA Safety Standards Series* (RS-G-1.2).

ICRP (1994). 'ICRP Publication 66: Human respiratory tract model for radiological protection'. In: *Annals of the ICRP* **24** (1-3). DOI: 10.1016/0146-6453(94)90029-9.

ICRP (1997). 'ICRP Publication 78: Individual monitoring for internal exposure of workers'. In: *Annals of the ICRP* **27** (3-4): 1–161. DOI: 10.1016/S0146-6453(98)00004-9.

ICRP (2002). 'ICRP Publication 89: Basic anatomical and physiological data for use in radiological protection: reference values'. In: *Annals of the ICRP* **32** (3-4). DOI: 10.1016/S0146-6453(03)00002-2.

ICRP (2007). 'ICRP Publication 103: The 2007 recommendations of the international commission on radiological protection'. In: *Annals of the ICRP* **37** (2-4). DOI: 10.1016/j.icrp.2007.11.001.

ICRP (2009). 'ICRP Publication 110: Adult reference computational phantoms'. In: *Annals of the ICRP* **39** (2). DOI: 10.1016/j.icrp.2009.07.004.

ICRP (2010). 'ICRP Publication 116: Conversion coefficients for radiological protection quantities for external radiation exposures'. In: *Annals of the ICRP* **40** (2-5). DOI: 10.1016/j.icrp.2011.10.001.

ICRU (1992a). 'ICRU Report 46: Photon, electron, proton and neutron interaction data for body tissues'. In: *Journal of the ICRU*.

ICRU (1992b). 'ICRU Report 48: Phantoms and computational models in therapy, diagnosis and protection'. In: *Journal of the ICRU*.

ICRU (2003). 'ICRU Report 69: Direct determination of the body content of radionuclides'. In: *Journal of the ICRU* **3** (1).

ICRU (2008). 'ICRU Report 80: Dosimetry systems for use in radiation processing'. In: *Journal of the ICRU* **8** (2).

ICRU (2011a). 'ICRU Report 85: Fundamental quantities and units for ionizing radiation'. In: *Journal of the ICRU* **11** (1).

ICRU (2011b). 'ICRU Report 86: Quantification and reporting of low-dose and other heterogeneous exposures'. In: *Journal of the ICRU* **11** (2).

INGBER, L. (1989). 'Very fast simulated re-annealing'. In: *Mathematical and Computer Modelling* **12** (8): 967–973. DOI: 10.1016/0895-7177(89)90202-1.

INTERNATIONAL ORGANIZATION FOR STANDARDIZATION (2010b). *Radiation protection — Performance criteria for radiobioassay*. Tech. rep. ISO/FDIS 28218:2010(E). Geneva, Switzerland: International Organization for Standardization.

INTERNATIONAL ORGANIZATION FOR STANDARDIZATION (2010a). *Determination of the characteristic limits (decision threshold, detection limit and limits of the confidence interval) for measurements of ionizing radiation — Fundamentals and application.* Tech. rep. ISO 11929:2010. Geneva, Switzerland: International Organization for Standardization.

JAVA COMMUNITY (2012). *JAXB reference implementation.* URL: http://jaxb.java.net.

JOHNSON, P. B., S. R. WHALEN, M. WAYSON, B. JUNEJA, C. LEE and W. E. BOLCH (2009). 'Hybrid patient-dependent phantoms covering statistical distributions of body morphometry in the U.S. adult and pediatric population'. In: *Proceedings of the IEEE* **97** (12): 2060–2075. DOI: 10.1109/JPROC.2009.2032855.

JU, T., F. LOSASSO, S. SCHAEFER and J. WARREN (2002). 'Dual contouring of hermite data'. In: *ACM Transactions on Graphics* **21** (3): 339–346. DOI: 10.1145/566654.566586.

KAWRAKOW, I., E. MAINEGRA-HING and D. W. O. ROGERS (2006). *EGSnrcMP: The multi-platform environment for EGSnrc.* Tech. rep. NRCC PIRS-877. Ottawa, Canada: Ionizing Radiation Standards, National Research Council of Canada.

KAWRAKOW, I., E. MAINEGRA-HING, D. W. O. ROGERS, F. TESSIER and B. R. B. WALTERS (2011). *The EGSnrc code system: Monte Carlo simulation of electron and photon transport.* Tech. rep. NRCC PIRS-701. Ottawa, Canada: Ionizing Radiation Standards, National Research Council of Canada.

KENDALL, M. G. and J. D. GIBBONS (1990). *Rank correlation methods.* 5th ed. London and New York, NY: Edward Arnold.

KIM, C. H., J. H. JEONG, W. E. BOLCH, K.-W. CHO and S. B. HWANG (2011). 'A polygon-surface reference Korean male phantom (PSRK-Man) and its direct implementation in Geant4 Monte Carlo simulation'. In: *Physics in Medicine and Biology* **56** (10): 3137–3161. DOI: 10.1088/0031-9155/56/10/016.

'Advanced Monte Carlo for radiation physics, particle transport simulation and applications' (2001). In: *Proceedings of the Monte Carlo 2000 Conference, Lisbon, 23-26 October 2000.* Ed. by A. KLING, F. BARAO, M. NAKAGAWA, L. TAVORA and P. VAZ. Springer-Verlag Berlin Heidelberg New York.

KNOLL, G. F. (2010). *Radiation detection and measurement.* 4th ed. John Wiley & Sons.

KOHAVI, R. and G. H. JOHN (1997). 'Wrappers for feature subset selection'. In: *Artificial Intelligence* **97** (1-2): 273–324. DOI: 10.1016/S0004-3702(97)00043-X.

KOROLIUK, V. S. (2013). 'B.V. Gnedenko: Classic of limit theorems in the theory of probability'. In: *Methodology and Computing in Applied Probability*. DOI: `10.1007/s11009-013-9353-8`.

KRAMER, G. H., L. C. BURNS and S. YIU (1997). 'Lung counting: Evaluation of uncertainties in lung burden estimation arising from a heterogeneous lung deposition using Monte Carlo code simulations'. In: *Radiation Protection Dosimetry* **74** (3): 173–182. DOI: `10.1093/oxfordjournals.rpd.a032194`.

KRAMER, G. H., B. M. HAUCK and S. A. ALLEN (2001). 'Chest wall thickness measurements and the dosimetric implications for male workers in the uranium industry'. In: *Health Physics* **80** (1): 74–80.

KRAMER, G. H. (2007). 'The efficiency curve: a new function'. In: *Radiation Protection Dosimetry* **127** (1-4): 270–272. DOI: `10.1093/rpd/ncm348`.

LABORATOIRE NATIONAL HENRI BECQUEREL (2013). *Nucleide gamma and alpha library*. Gif-sur-Yvette Cedex, France. URL: `www.nucleide.org/DDEP_WG/DDEPdata.htm`.

LAUBERSHEIMER, S. (2011). 'Segmentierung von CT-Aufnahmen des JAERI-Phantoms zwecks Vorbereitung von numerischen Zähleffizienzberechnungen in der in vivo Messtechnik'. Student thesis. Karlsruhe Institute of Technology.

LAUBERSHEIMER, S. (2012). 'Simulationsbasierte Kalibrierung eines Teilkörperzählers mit Reinstgermaniumdetektoren für den Nachweis inkorporierter Radionuklide in Teilkörpergeometrien und zur Messung von Radionukliden in beliebigen Proben'. Diploma thesis. Karlsruhe Institute of Technology.

LEE, C., C. LEE, D. LODWICK and W. E. BOLCH (2007). 'Hybrid computational phantoms of the male and female newborn patient: NURBS-based whole-body models'. In: *Physics in Medicine and Biology* **52** (12): 3309–3333. DOI: `10.1088/0031-9155/52/12/001`.

LEE, C., D. LODWICK, J. HURTADO, D. PAFUNDI, J. L. WILLIAMS and W. E. BOLCH (2010). 'The UF family of reference hybrid phantoms for computational radiation dosimetry'. In: *Physics in Medicine and Biology* **55** (2): 339–363. DOI: `10.1088/0031-9155/55/2/002`.

LENNERZ, C. and E. SCHÖMER (2002). 'Efficient distance computation for quadratic curves and surfaces'. In: *Geometric Modeling and Processing*: 60–69. DOI: `10.1109/GMAP.2002.1027497`.

LIYE, L., D. FRANCK, L. DE CARLAN and L. JUNLI (2007). 'Application of Monte Carlo calculation and OEDIPE software for virtual calibration of an in vivo counting system'. In: *Radiation Protection Dosimetry* **127** (1-4): 282–286. DOI: `10.1093/rpd/ncm469`.

LOPEZ, M. A., I. BALÁSHÁZY, P. BÉRARD, E. BLANCHARDON, B. BREUSTEDT, D. BROGGIO, C. M. CASTELLANI, D. FRANCK, A. GIUSSANI, C. HURTGEN, A. C. JAMES, W. KLEIN, G. H. KRAMER, W. B. LI, J. W. MARSH, I. MALATOVA, D. NOSSKE, U. OEH, G. PAN, M. PUNCHER, P. PEIXOTO TELLES, P. TEIXOTO TELLES, J. SCHIMMELPFENG and T. VRBA (2011). 'EURADOS coordinated action on research, quality assurance and training of internal dose assessments'. In: *Radiation Protection Dosimetry* **144** (1-4): 349–352. DOI: 10.1093/rpd/ncq435.

LORENSEN, W. E. and H. E. CLINE (1987). 'Marching cubes: A high resolution 3D surface construction algorithm'. In: *ACM SIGGRAPH Computer Graphics* **21** (4): 163–169. DOI: 10.1145/37402.37422.

LYNCH, T. P. (2011). *In vivo monitoring program manual, PNL-MA-754*. Tech. rep. PNNL-19516. Richland, WA: Pacific Northwest National Laboratory.

MACKAY, D. J. C. (2003). *Information theory, inference and learning algorithms*. Cambridge University Press.

MARZOCCHI, O., B. BREUSTEDT and M. URBAN (2010). 'Characterisation, modelling and optimisation of the model of a HPGe detector with the aid of point sources'. In: *Applied Radiation and Isotopes* **68** (7-8): 1438–1440. DOI: 10.1016/j.apradiso.2009.11.022.

MARZOCCHI, O., B. BREUSTEDT, D. MOSTACCI and M. URBAN (2011). 'Comparison of stretched and sitting configurations for partial-body measurements'. In: *Applied Radiation and Isotopes: Including Data, Instrumentation and Methods for Use in Agriculture, Industry and Medicine* **69** (8): 1156–1158. DOI: 10.1016/j.apradiso.2010.11.012.

MARZOCCHI, O. (2011). 'Design and setup of a new HPGe detector based body counter capable of detecting also low energy photon emitters'. PhD thesis. University of Bologna.

MATERIALISE NV (2013). *Mimics*. Leuven, Belgium. URL: http://biomedical. materialise.com/mimics.

MCNAMARA, A. L., H. HEIJNIS, D. FIERRO and M. I. REINHARD (2012). 'The determination of the efficiency of a Compton suppressed HPGe detector using Monte Carlo simulations'. In: *Journal of Environmental Radioactivity* **106**: 1–7. DOI: 10.1016/j.jenvrad.2011.10.017.

METROPOLIS, N. and S. ULAM (1949). 'The Monte Carlo method'. In: *Journal of the American Statistical Association* **44** (247): 335–341. DOI: 10.1080/01621459.1949.10483310.

MIERSWA, I., M. WURST, R. KLINKENBERG, M. SCHOLZ and T. EULER (2006). 'YALE: Rapid prototyping for complex data mining tasks'. In: *Proceedings of the 12th ACM SIGKDD international conference on Knowledge discovery and data mining — KDD '06.* Ed. by L. UNGAR, M. CRAVEN, D. GUNOPULOS and T. ELIASSI-RAD. New York, New York, USA: ACM Press: 935–940. DOI: 10.1145/1150402.1150531.

MOFRAD, F. B., R. A. ZOROOFI, A. A. TEHRANI-FARD, S. AKHLAGHPOOR, M. HORI, Y.-W. CHEN and Y. SATO (2010). 'Statistical construction of a Japanese male liver phantom for internal radionuclide dosimetry'. In: *Radiation Protection Dosimetry* **141** (2): 140–148. DOI: 10.1093/rpd/ncq164.

MOHR, U. and B. BREUSTEDT (2007). *Messung von inkorporierten Radionukliden mittels Gammaspektrometrie im Teilkörperzähler mit Phoswich-Detektor.* Tech. rep. MB HS 012. Hauptabteilung Sicherheit, Forschungszentrum Karlsruhe.

MORTENSON, M. E. (1985). *Geometric modeling.* 3rd ed. South Norwalk, CT: Industrial Press Inc.

MUKHERJEE, S., P. TAMAYO, D. SLONIM, A. VERRI, T. GOLUB, J. P. MESIROV and T. POGGIO (1999). *Support vector machine classification of microarray data.* Tech. rep. Artificial Intelligence Laboratory, Massachusetts Institute of Technology. URL: http://cbcl.csail.mit.edu/projects/cbcl/res-area/abstracts/2001-abstracts/mukherjee/microarray.ps.

NA, Y. H., B. ZHANG, J. ZHANG, P. F. CARACAPPA and X. G. XU (2010). 'Deformable adult human phantoms for radiation protection dosimetry: Anthropometric data representing size distributions of adult worker populations and software algorithms'. In: *Physics in Medicine and Biology* **55** (13): 3789–3811. DOI: 10.1088/0031-9155/55/13/015.

NAGAOKA, T. and S. WATANABE (2008). 'Postured voxel-based human models for electromagnetic dosimetry'. In: *Physics in Medicine and Biology* **53** (24): 7047–7061. DOI: 10.1088/0031-9155/53/24/003.

NATIONAL INFORMATION STANDARDS ORGANIZATION (2005). *Guidelines for the construction, format, and management of monolingual controlled vocabularies.* Tech. rep. ANSI/NISO Z39.19. Baltimore, MD: National Information Standards Organization.

NATIONAL NUCLEAR DATA CENTER (2013a). 'Evaluated Nuclear Structure Data File retrieval'. In: URL: www.nndc.bnl.gov/ensdf.

NATIONAL NUCLEAR DATA CENTER (2013b). *NuDat 2.6.* Upton, NY. URL: www.nndc.bnl.gov/nudat2.

NIELSON, G. M. and B. HAMANN (1991). 'The asymptotic decider: Resolving the ambiguity in marching cubes'. In: *Proceedings of the 2nd Conference on Visualization '91*. Los Alamitos, CA, USA: IEEE Computer Society Press: 83–91.

NOGUEIRA, P., L. SILVA, P. TELES, J. BENTO and P. VAZ (2010). 'Monte Carlo simulation of the full energy peak efficiency of a WBC'. In: *Applied Radiation and Isotopes: Including Data, Instrumentation and Methods for Use in Agriculture, Industry and Medicine* **68** (1): 184–189. DOI: `10.1016/j.apradiso.2009.09.014`.

PELLED, O., U. GERMAN, G. POLLAK and Z. ALFASSI (2006). 'Estimation of errors due to inhomogeneous distribution of radionuclides in lungs'. In: *Nuclear Instruments and Methods in Physics Research Section A: Accelerators, Spectrometers, Detectors and Associated Equipment* **564** (1): 491–495. DOI: `10.1016/j.nima.2006.04.036`.

PELOWITZ, D. B. (2007). *MCNPX user's manual*. Tech. rep. LA-CP-07-1473. Los Alamos, NM: Los Alamos National Laboratory.

PIEGL, L. A. and W. TILLER (1997). *The NURBS book*. 2nd ed. Springer-Verlag Berlin Heidelberg New York.

PIERRAT, N., G. PRULHIERE, L. DE CARLAN and D. FRANCK (2007). 'Determination of new European biometric equations for the calibration of in vivo lung counting systems using the Livermore phantom'. In: *Radiation Protection Dosimetry* **125** (1-4): 449–455. DOI: `10.1093/rpd/ncm150`.

PÖLZ, S., T. SCHNEIDER, L. HEGENBART, M. URBAN and H. GECKEIS (2012). 'Quantification of the effect of respiratory motion on efficiency calibration for internal dosimetry'. In: *The 13th International Congress of the International Radiation Protection Association, 13-18 May 2012*. Glasgow, Scotland.

PÖLZ, S., S. LAUBERSHEIMER, J. S. EBERHARDT, M. A. HARRENDORF, T. KECK, A. BENZLER and B. BREUSTEDT (2013). 'Voxel2MCNP: A framework for modeling, simulation and evaluation of radiation transport scenarios for Monte Carlo codes'. In: *Physics in Medicine and Biology* **58** (16): 5381–5400. DOI: `10.1088/0031-9155/58/16/5381`.

QT PROJECT and DIGIA (2013). *Qt*. URL: `http://qt-project.org`.

REILLY, D., N. ENSSLIN, H. SMITH, JR. and S. KREINER (1991). *Passive nondestructive assay of nuclear materials*. Tech. rep. NUREG/CR-5550. Springfield, VA: Los Alamos National Laboratory. URL: `www.lanl.gov/orgs/n/n1/panda`.

RIPER, K. A. V. (2003a). '3D display of very large MCNPX and MCNP lattices in Moritz'. In: *Proceedings of the Nuclear Mathematical and Computational*

Sciences: A Century in Review, A Century Anew, Gatlinburg, Tennessee, April 6-11, 2003: on CD–ROM.

RIPER, K. A. V. (2003b). 'Analysis of and refinements to the Sabrina volume fraction algorithm'. In: *Proceedings of the Nuclear Mathematical and Computational Sciences: A Century in Review, A Century Anew, Gatlinburg, Tennessee, April 6-11, 2003*: on CD–ROM.

RISSANEN, J. (1978). 'Modeling by shortest data description'. In: *Automatica* **14**. DOI: `10.1016/0005-1098(78)90005-5`.

ROBERT MCNEEL & ASSOCIATES (2012). *Rhinoceros*. Seattle, WA. URL: `www.rhino3d.com`.

ROSSET, A., L. SPADOLA and O. RATIB (2004). 'OsiriX: An open-source software for navigating in multidimensional DICOM images'. In: *Journal of Digital Imaging* **17** (3): 205–216. DOI: `10.1007/s10278-004-1014-6`.

SALVAT, F., J. M. FERNÁNDEZ-VAREA and J. SEMPAU (2011). *PENELOPE-2011: A code system for Monte Carlo simulation of electron and photon transport*. Tech. rep. OECD NEA Data Bank/NSC DOC(2011)/5. Issy-les-Moulineaux, France: OECD Nuclear Energy Agency.

SCHAEFER, S., T. JU and J. WARREN (2007). 'Manifold dual contouring'. In: *IEEE Transactions on Visualization and Computer Graphics* **13** (3): 610–619. DOI: `10.1109/TVCG.2007.1012`.

SCHAPIRE, R. E. (2003). 'The boosting approach to machine learning: An overview'. In: *Nonlinear Estimation and Classification*: 1–23. URL: `https://www.cs.princeton.edu/courses/archive/spring07/cos424/papers/boosting-survey.pdf`.

SCHNEIDER, T. (2011a). 'Auswirkung der Atembewegung auf die Zähleffizienz von in vivo Messsystemen bei Lungenteilkörpermessungen'. Bachelor thesis. Duale Hochschule Baden-Württemberg Karlsruhe.

SCHNEIDER, T. (2011b). 'Erstellung von Voxelmodellen zur Bestimmung der Atembewegungsabhängigkeit der Zähleffizienz bei in vivo Inkorporationsmessungen'. Student thesis. Duale Hochschule Baden-Württemberg Karlsruhe.

SCHOWE, B. (2010). *Feature selection for high-dimensional data with RapidMiner*. Tech. rep. Technical University of Dortmund. URL: `www-ai.cs.uni-dortmund.de/PublicPublicationFiles/schowe_2011a.pdf`.

SCHWARZ, R. (2008). *Graphical user interface for high energy multi-particle transport*. Tech. rep. November. Richland, WA: Visual Editor Consultants.

SEGARS, W. P. and G. M. STURGEON (2010). 'The new XCAT series of digital phantoms for multi-modality imaging'. In: *Nuclear Science Symposium Conference Record (NSS/MIC), 2010 IEEE*: 2392–2395.

SEGARS, W. P. and B. M. W. TSUI (2009). 'MCAT to XCAT: The evolution of 4-D computerized phantoms for imaging research'. In: *Proceedings of the IEEE* **97** (12): 1954–1968. DOI: 10.1109/JPROC.2009.2022417.

SHIROTANI, T. (1988). 'Realistic torso phantom for calibration of in-vivo transuranic-nuclide counting facilities'. In: *Journal of Nuclear Science and Technology* **25** (11): 875–883. DOI: 10.1080/18811248.1988.9735941.

SHREINER, D. (1999). *OpenGL reference manual: The official reference document to OpenGL.* 3rd ed. Boston, MA: Addison-Wesley Longman Publishing Co., Inc.

SHULTIS, J. K. and R. E. FAW (2006). *An MCNP primer.* Tech. rep. c. Manhattan, KS: Kansas State University.

STICHTING BLENDER FOUNDATION (2013). *Blender.* Amsterdam, the Netherlands. URL: www.blender.org.

SUMERLING, T. J. and S. P. QUANT (1982). 'Measurements of the human anterior chest wall by ultrasound and estimates of chest wall thickness for use in determination of transuranic nuclides in the lung'. In: *Radiation Protection Dosimetry* **3** (4): 203–210.

THE R CORE TEAM (2013). *R: A language and environment for statistical computing.* Tech. rep. Vienna, Austria: R Foundation for Statistical Computing. URL: www.r-project.org.

THEIS, C., K. H. BUCHEGGER, M. BRUGGER, D. FORKEL-WIRTH, S. ROESLER and H. VINCKE (2006). 'Interactive three-dimensional visualization and creation of geometries for Monte Carlo calculations'. In: *Nuclear Instruments and Methods in Physics Research Section A* **562** (2): 827–829. DOI: 10.1016/j.nima.2006.02.125.

TRAUB, R. J. (2008). *Influence of manufacturing processes on the performance of phantom lungs.* Tech. rep. PNNL-17852. Richland, WA: Pacific Northwest National Laboratory.

TULI, J. K. (2001). *Evaluated nuclear structure data file: A manual for preparation of data sets.* Tech. rep. BNL-NCS-51655-01/02-Rev. Upton, NY: National Nuclear Data Center, Brookhaven National Laboratory.

TWARD, D. J., C. CERITOGLU, A. KOLASNY, G. M. STURGEON, W. P. SEGARS, M. I. MILLER and J. T. RATNANATHER (2011). 'Patient specific dosimetry phantoms using multichannel LDDMM of the whole body'. In: *International Journal of Biomedical Imaging* **2011**: 1–9. DOI: 10.1155/2011/481064.

U.S. NATIONAL LIBRARY OF MEDICINE (2012). *Medical subject headings.* URL: www.nlm.nih.gov/mesh.

VAPNIK, V. N. (1999). 'An overview of statistical learning theory'. In: *IEEE transactions on neural networks* **10** (5): 988–999. DOI: 10.1109/72.788640.

VAPNIK, V. N. and A. Y. CHERVONENKIS (1971). 'On the uniform convergence of relative frequencies of events to their probabilities'. In: *Theory of Probability and its Applications* **16** (2): 264–280. DOI: 10.1137/1116025.

VEILLARD, D. (2012). *Libxml2: The XML C parser and toolkit of Gnome*. URL: www.xmlsoft.org.

W3C (2004). *XML Schema 1.1*. Tech. rep. URL: www.w3.org/XML/Schema.

WESTON, J. and A. ELISSEEFF (2003). 'Use of the zero norm with linear models and kernel methods'. In: *The Journal of Machine Learning Research* **3**: 1439–1461. URL: http://dl.acm.org/citation.cfm?id=944982.

XU, X. and K. ECKERMAN, eds. (2009). *Handbook of anatomical models for radiation dosimetry*. Boca Raton, FL: Taylor & Francis.

YANG, A. Y., J. WRIGHT, Y. MA and S. S. SASTRY (2007). *Feature selection in face recognition: A sparse representation perspective*. Tech. rep. UCB/EECS-2007-99. Berkeley, CA: EECS Department, University of California. URL: http://citeseerx.ist.psu.edu/viewdoc/download?doi=10.1.1.112.6819&rep=rep1&type=pdf.

YU, C. H. (2003). 'Resampling methods: Concepts, applications, and justification'. In: *Practical Assessment, Research & Evaluation* **8** (19): 1–20. URL: http://pareonline.net/getvn.asp?v=8&n=19.

ZAIDI, H. and B. M. W. TSUI (2009). 'Review of computational anthropomorphic anatomical and physiological models'. In: *Proceedings of the IEEE* **97** (12): 1938–1953. DOI: 10.1109/JPROC.2009.2032852.

ZAIDI, H. and X. G. XU (2007). 'Computational anthropomorphic models of the human anatomy: The path to realistic Monte Carlo modeling in radiological sciences'. In: *Annual Review of Biomedical Engineering* **9**: 471–500. DOI: 10.1146/annurev.bioeng.9.060906.151934.

ZHANG, J., Y. H. NA, P. F. CARACAPPA and X. G. XU (2009). 'RPI-AM and RPI-AF, a pair of mesh-based, size-adjustable adult male and female computational phantoms using ICRP-89 parameters and their calculations for organ doses from monoenergetic photon beams'. In: *Physics in Medicine and Biology* **54** (19): 5885–5908. DOI: 10.1088/0031-9155/54/19/015.